Risk Engineering

Series editor

Dirk Proske, Vienna, Austria

The Springer Book Series *Risk Engineering* can be considered as a starting point, looking from different views at Risks in Science, Engineering and Society. The book series publishes intense and detailed discussions of the various types of risks, causalities and risk assessment procedures.

Although the book series is rooted in engineering, it goes beyond the thematic limitation, since decisions related to risks are never based on technical information alone. Therefore issues of "perceived safety and security" or "risk judgment" are compulsory when discussing technical risks, natural hazards, (environmental) health and social risks. One may argue that social risks are not related to technical risks, however it is well known that social risks are the highest risks for humans and are therefore immanent in all risk trade-offs. The book series tries to cover the discussion of all aspects of risks, hereby crossing the borders of scientific areas.

More information about this series at http://www.springer.com/series/11582

Dirk Proske

Bridge Collapse Frequencies versus Failure Probabilities

With 29 Tables and 90 Figures

Springer

Dirk Proske
University of Natural Resources and
Life Sciences
Vienna
Austria

ISSN 2195-433X ISSN 2195-4348 (electronic)
Risk Engineering
ISBN 978-3-030-08864-4 ISBN 978-3-319-73833-8 (eBook)
https://doi.org/10.1007/978-3-319-73833-8

Printed on acid-free paper

This Springer imprint is published by the registered company Springer International Publishing AG part of Springer Nature
The registered company address is: Gewerbestrasse 11, 6330 Cham, Switzerland

For Beatrix and Rolf

Contents

Chapter 1
Objective

Probably more than a billion structures exist on earth including several million bridges. The success of the technical product "structures" is not only based on the gained large improvement of the quality of life for humans including protection against environmental hazards and conditions and security, it is also strongly related to the outstanding safety of the structures itself.

Structures are probably one of the earliest technical products produced by mankind (Figs. 1.1 and 1.2). The code of Hammurabi by imposing harsh punishment to builders of collapsing structures shows that the safety of structures has been an important issue since thousands of years. The tools to ensure and provide a sufficient safety of structures have evolved over this time, for arch bridges see Proske and van Gelder (2009).

Today there exist different numerical parameters to evaluate the safety of structures. One of these parameters is the "probability of failure" of a structure. The term probability of failure and failure probability of structures respectively can be found in a large number of scientific publications, books or codes of practice. The computation of the probability of failure of structures, and specifically bridges, is perhaps not daily business, but state-of-the-art and has been carried out in numerous cases [just see the conference proceedings of the International Conference on Structural Safety & Reliability (ICOSSAR), International Conference on Applications of Statistics and Probability in Civil Engineering (ICASP), European Safety and Reliability Conference (ESREL) and International Probabilistic Workshop series (IPW)]. If we have computed the probability of failure in so many cases we should be able to compare these theoretical values with the observation, the collapse frequency.

However, such comparisons of the probability of failure with the frequency of collapse are not common in structural engineering, they do not exist for bridges. In many codes and books it is argued that these two parameters can not directly be compared due to their individual limitations. This argument is surprising since in other industries such as the Nuclear Power Industry such comparisons are carried out (Proske 2016) and are often an issue of public discussion related to the quality of the models. Even further, our models should always be comparable to reality and to real

D. Proske, *Bridge Collapse Frequencies versus Failure Probabilities*, Risk Engineering,
https://doi.org/10.1007/978-3-319-73833-8_1

Fig. 1.1 Reconstruction of a typical pile dwelling around 4000 B.C in Switzerland (Seengen) (*Picture* D. Proske)

Fig. 1.2 Giza pyramid complex build around 2500 B.C. (*Picture* D. Proske)

world observations respectively which would in our case be the frequency of collapse. If our models are neither based, nor compared nor confirmed by observations, we develop models and parameters which are neither useful nor verifiable. We have to discard them or to enhance the theoretical parameters to parameters which are useful and verifiable, in other terms which can be related to real-word measurements or observations.

Therefore the objective of this book is the comparison of the computed probability of failure of bridges with the observed frequency of collapse of bridges. To carry out this objective, in this book techniques are applied to resolve the limitations of both parameters to gain comparability. These extensions are for example related to the issues of human failure and to the consideration of correlation of the individual limit states to the structural value.

Additionally the time dependency of the probability of failure and the collapse frequency is considered and prediction of future developments are attempted. This is of utmost importance since the maintenance of the large stock of bridges in developed and developing countries has to be carried out by limited resources. Many conferences deal with this issue, see for example the proceedings of the IALCCE-conferences or the Dresden Bridge Symposium.

Finally the available data of the frequency of collapse is studied in detail to recognize patters and to identify potential weakness in our design and in our maintenance strategies. This conclusion may affect the future development of codes of practice as well as enable the clients and the authorities to focus their attention.

By doing so, we act responsible in terms of safety and use of our resources. One thing engineers have to keep in mind: we always can extend our knowledge and we must learn from the past to develop strategies for a better future. This can only be done if we accept the criterion of practice as final criteria for the objective truth.

Beside all the considerations above related to the probability of failure and the collapse frequency, several authors have compared the risk of bridge failure to overall risk measures and have hereby shown that structures (Proske 2009; Tanner and Hingorani 2015) in general and specifically bridges (Curbach et al. 2002; Imhof 2004) are a very safe technical product. This is a strong indirect indication that the safety concept based on probabilities of failure functions.

References

Curbach M, Nitzsche W-M, Proske D (2002) Die Sicherheit von Brücken im Vergleich zu anderen Risiken, Tagungsband 12. Dresdner Brückenbausymposium, 14. März 2002, TU Dresden, pp 197–218

Imhof D (2004) Risk assessment of existing bridge structures. Ph.D. Dissertation, University of Cambridge, Kings College

Proske D (2009) Catalogue of risk. Springer, Berlin, Heidelberg

Proske D, van Gelder P (2009) Safety of historical stone arch bridges. Springer, Berlin, Heidelberg

Proske D (2016) Differences between probability of failure and probability of core damage. In: 14th international probabilistic workshop, Ghent, Springer, Cham, pp 109–122

Tanner P, Hingorani R (2015) Acceptable risks to persons associated with building structures. Struct Concr 3:314–322

Chapter 2
Terms and Definitions

2.1 Introduction

To provide the reader a clear understanding, this chapter introduces definitions of the most important words and terms used in the book. The chapter does not include all terms since this would exceed the volume of the book.

In Proske (2009) the author has already showed the general objective and basis of the definition of terms: The goal of the introduction and application of terms is to boost communication. A term contains all properties which belong to a thinking unit (DIN 2330 1979). The definition is the assignment of the content to a term. A definition of a term should be true, useful and fundamental. However, all terms include objective parts (denotation) and subjective parts (connotation), which limit the goals of the term, e.g. common understanding. For example, a term may be clearly distinguishing some properties, for other properties only rough distinction can be made. This is shown in Fig. 2.1. Also definitions can be time-dependent.

2.2 Definition of the Term "Bridge"

According to the Association of American State Highway and Transportation Officials (AASHTO) the term bridge is defined as "a structure, including supports, erected over a depression or an obstruction (such as water, highway and railway) having a track or passage way for carrying traffic or other moving loads." (taken from Hersi 2009; Imhof 2004).

A South-African definition of a bridge states: "A bridge is a structure erected with a deck for carrying traffic over or under an obstruction and with a clear span of 6 m or more. Where the clear span is less than 6 m, reference is to a culvert" (Wolhuter 2015).

D. Proske, *Bridge Collapse Frequencies versus Failure Probabilities*, Risk Engineering,
https://doi.org/10.1007/978-3-319-73833-8_2

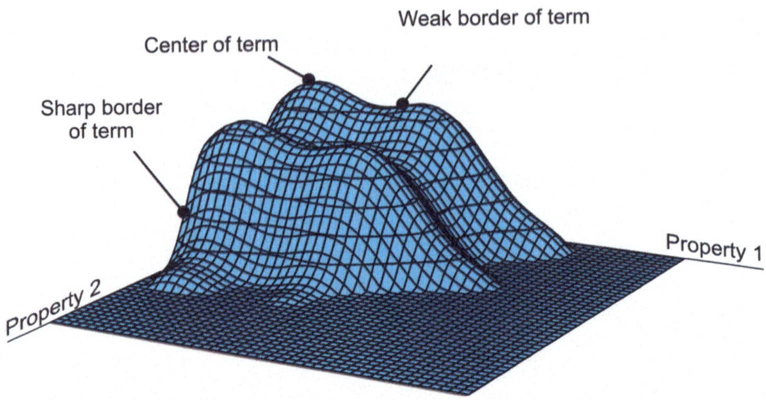

Fig. 2.1 Landscape of a term fitting to some properties (Proske 2009; Riedl 2000)

The National Bridge Inspection Standards give the following definition: "A structure including supports erected over a depression or an obstruction, such as water, highway, or railway, and having a track or passageway for carrying traffic or other moving loads, and having an opening measured along the centre of the roadway of more than 20 ft (6.1 m) between undercopings of abutments or spring lines of arches, or extreme ends of openings for multiple boxes; it may also include multiple pipes, where the clear distance between openings is less than half of the smaller contiguous opening" (taken from FHWA 1996).

A German definition considers bridges as all transitions of a traffic route over another traffic route, over a body of water or over a lower ground, if the clearance between the abutments is 2.00 m or more (DIN 1076).

It can be seen that the threshold differs between different regions, which limits the comparison of data from different regions.

2.3 Definition of the Term "Collapse"

The author considers bridge collapse as a state in which bridge parts are no longer held in position by the structure itself. Further definitions are given below.

A bridge collapse occurs when an entire bridge or a substantial part comes down, at which point the structure loses its ability to perform its function (Wardhana and Hadipriono 2003).

Another definition sees collapse as the development of a cinematic chain, which yields to total or partial destruction of the bridge. A local cross section failure or the exceedance of certain limit state values does not necessarily yield to a collapse since the bridge may remain in a condition in which it can be repaired (Klingmüller and Bourgund 1992).

According to Imhof (2004) a "…collapse of a significant part of the structure or the complete structure, both of which render the structure incapable of remaining in service."

The New York State Department of Transportation (NYSDOT 2004, Cook 2014) has introduced two definitions for collapse: total and partial collapse. A total collapse is defined as "structures which all primary members of a span or several spans have undergone severe deformation such that no travel lanes are passable." A partial collapse is defined as "structures on which all or some of the primary structural members of a span or multiple spans have undergone severe deformation such that the lives of those travelling on or under the structure would be in danger".

In contrast, for seismic loads several codes define a state called "repairable damage" limit state. At this state there is no noticeable risk of life (Gkatzogias and Kappos 2015).

2.4 Definition of the Term "Cause"

The definition of a cause of a bridge collapse is very ambiguous. One possible technique is root cause analysis (RCA), a technique to identify the root causes of a problem. A root cause is a factor whose removal from the sequence will prevent the problem whereas a causal factor only affects the scale of the problem.

The related field of science to identify the human cause of an event is the jurisprudence. However, such a discussion including trials, judgements and many expert assessments goes beyond the goal of the book. As an example Fig. 2.2 shows the perception of causes of disasters. The picture shows that even disasters strongly related to extreme natural hazards such as earthquakes, floods and tornados include partial man-made contributions such as collapsed bridges. Common definitions of causes such as shown in Table 2.1 try to cover this problem by introducing classes such as limited knowledge. An alternative classification of causes is found in Vogel et al. (2009).

On the other hand, some authors simply declare the cause of collapse of bridges as a failure of the engineers. For example, Calvert (2002) stated that "bad design does not only mean errors of computation, but a failure to take into account the loads the structure will be called upon to carry, erroneous theories, reliance on inaccurate data, ignorance of the effects of repeated or impulsive stresses, and improper choice of materials or misunderstanding of their properties. The engineer is responsible for these failures, which are created at the drawing board."

The following statement with more or less the same intention is found in Taricska (2014) and Chavel and Yadlosky (2011): Principal causes are defined as "errors in design, detailing, or construction; unanticipated effects of stress concentrations; lack of proper maintenance; the use of improper materials or foundation type; or the insufficient consideration of an extreme event."

To avoid these extreme standpoints (e.g. 100% human failure versus 100% natural hazard) we find and use publications declaring primary (enabling cause), secondary

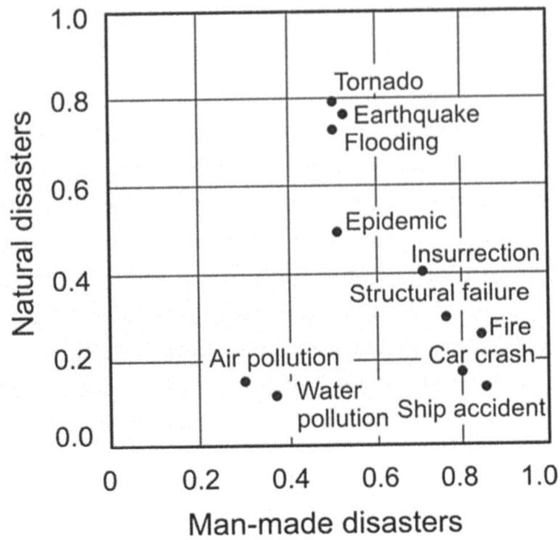

Fig. 2.2 Perception of causes of disasters (Karger 1996)

(triggering cause) and tertiary causes (management cause) (Hersi 2009; Taricska 2014). This approach is comparable, at least to a certain extend, to the strategies found in Nuclear Engineering, where usually several causes are considered, reaching from the initiating event such as an earthquake to management failure (loss of safety culture). The disadvantage is that human, social, management and cultural failures can not be covered by the same scientific tools. This fact is visualised in Figs. 2.3 and 2.4: for different types of systems different tools are used. On the one hand, simple systems can be treated by analytical techniques (Fig. 2.3). Applied sciences and structural engineering belongs to this group which is bordered by other systems as shown in Fig. 2.4. On the other hand, systems with higher complexity have to be modelled by other techniques, e.g. statistical techniques.

To avoid this discussion this book simply applied the causes given in the references by the other authors. There may be an inhomogeneity in the definitions, however if the results in terms of the frequency of causes to the overall number is mainly robust, a conclusion can be drawn.

2.5 Definition of the Term "Bridge Collapse Frequency"

The terms bridge collapse frequency and frequency of bridge collapses are used with the same meaning in this book. The frequency of bridge collapse (F_c) is the ratio of the number of collapsed bridges n_c compared to the bridge stock number n_b related to a certain time scale (usually one year) and a certain region or country:

$$F_c = \frac{n_c}{n_b}.$$

Table 2.1 Classification of failure causes (Imhof 2004)

Failure cause	Nature	Example
Limited knowledge	Possible failure mode unrecognised Unknown phenomena	Unknown problems of fatigue, brittle failure Unknown buckling problems
Natural hazard	Extreme conditions More extreme and frequent hazard occurrence than assumed	Wind Storm Flooding
Design error (human error during design stage)	Omission of load or load combination Wrong assumption in ground condition Inadequate design of scaffolding	Calculation errors Error in software Unfavourable geo-technical properties not detected
Overloading	Accidental overloading Loading increased with time Change of use without structural assessment	Illegal overweight Changes to legal limit Special heavy-weight transports
Impact	Impact of ships Impact of vehicles Impact of trains	Loss of ship control Loss of vehicle control Bridge bashing over-height vehicles
Human error (non-design)	Workman use wrong material Workman change original design Poor workmanship Inadequate maintenance action	Change of original construction sequence Stiffeners welded to wrong section Scaffolding dismantled too early
Vandalism	Fire Explosion	Deliberately set fire Terrorist acts
Deterioration	Corrosion of steel reinforcing bars Corrosion of pres-tressing cables Concrete deterioration Fatigue	Loss of resistance of steel bars or hangers Loss of bond in RC structures Alkali-silica reaction, Freeze-thaw action

2.6 Definition of the Term "Failure Probability"

The terms failure probability and probability of failure are used with the same meaning in this book. The probability of failure P_{f} as a measure for safety can be referred to one year or the life time of a structure in n years. The measure is based on limit state function $g(\mathbf{X})$ and input random variables x (see Fig. 2.5):

$$P_f = \int \cdots \int_{g(\mathbf{X}) \leq 0} f_x(x)\, dx$$

$$P_f(n) = 1 - (1 - P_f)^n.$$

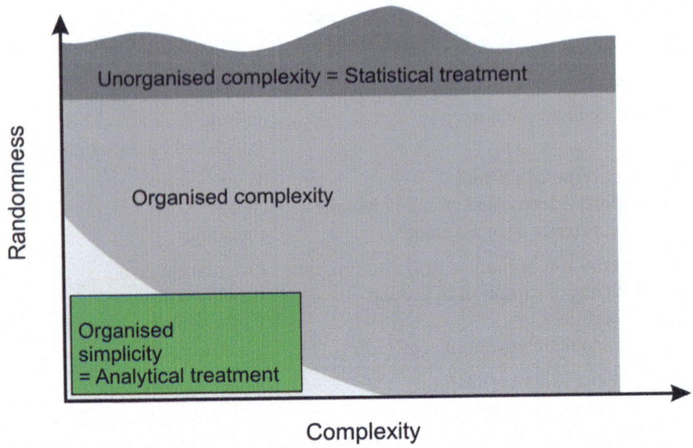

Fig. 2.3 Types of systems and the applied tools (Proske 2009; Weinberg 1975)

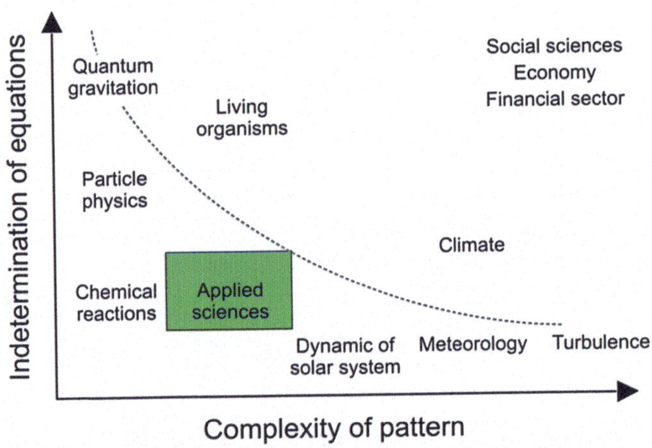

Fig. 2.4 Complexity of patterns and indetermination of associated equations for different systems (Proske 2009; Barrow 1998)

Since the probability of failure is extremely low due to the high safety requirements, a substitute measure, the safety index β, has been heavily applied instead of the probability in structural engineering. The safety index β is defined as:

$$P_f = \Phi(-\beta)$$

ϕ^{-1} represents the inverse standard normal distribution.

In general, the probability of failure P_f is not intended to be a true predicator of the collapse frequency of bridges. Many possible influence factors, such as human errors or correlations between different limit states seem to neither to be considered at all

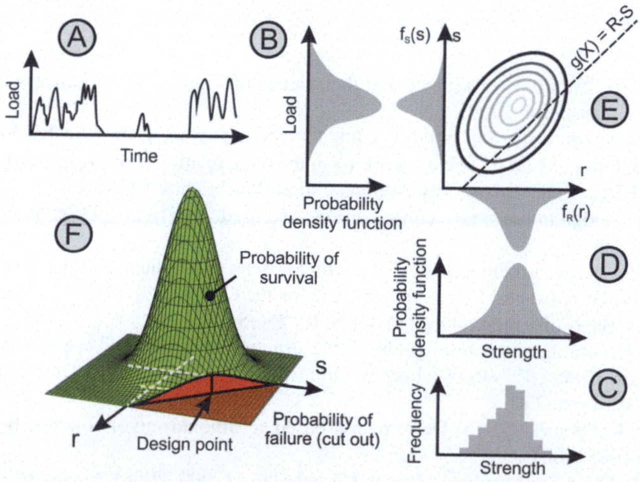

Fig. 2.5 Visualisation of the probability of failure as integral of a two-dimensional probability function including "a limit state. The statistical data of load and resistance variables are shown in charts A and C. The statistical evaluation is shown in charts B and D. The probability functions are combined with a limit state equation in chart E. In chart F, the two-dimensional joint probability function is represented three-dimensionally with the limit state equation and the design point

nor considered to an adequate extend. This is also clearly stated, for example see Table 2.1, Note 2 in the Eurocode 0 (EN 1990 2010). However, if the probability of failure is considered as an efficient, effective and robust safety measure, it must prove its worth in reality.

2.7 Conclusion

The most five important terms used in this book have been defined in this chapter (the big five). On the one hand, these definitions should enable the reader to understand the information and conclusions given in this document. On the other hand, the definitions also show the limitation of the comparability of the different studies since in different regions and different times different definitions were used. For example, in different countries a different minimum length for the definition of bridges is used. Although the effect of this may be limited to highway bridges, it limits the comparison of the entire bridge stock. The same conclusion can be drawn from the definition of the cause of a bridge collapse. Of course, a bridge collapse will always initiate juridical investigation which will define the cause of the collapse. Besides such individual investigations the systematic approach used in this document can not provide a homogenous procedure since different studies use different definitions of the cause. In this book the results regarding the definition of bridges and causes will be used from the study as they are.

References

Barrow JD (1998) The limits of science and the science of limits. Oxford University Press, New York (German translation 1999)
Calvert JB (2002) The collapse of buildings. https://mysite.du.edu/~jcalvert/tech/failure.htm
Chavel BW, Yadlosky JM (2011) Framework for improving resilience of bridge design. Tech. No. FHWA-IF-11-016. Federal Highway Administration, Washington DC
Cook W (2014) Bridge failure rates, consequences, and predictive trends. Dissertation, Utah State University Logan
DIN 1076 (1999) Ingenieurbauwerke im Zuge von Strassen und Brücken, November 1999
DIN 2330 (1979) Begriffe und Benennungen. Allgemeine Grundsätze
EN 1990 (2010) Basis of structural design, CEN/TC 250/SC 10
FHWA (1996) Recording and coding guide for the structure inventory and appraisal of the nations bridges. Report FHWA-PD-96-001 Federal Highway Administration, US Department of Transportation, Washington DC
Gkatzogias KI, Kappos AJ (2015) Deformation-based seismic design of concrete bridges. Earthq Structu 9(5):1045–1067
Hersi M (2009) Analysis of bridge failure in United States (2000–2008). Master of science thesis, Ohio State University
Imhof D (2004) Risk assessment of existing bridge structures. University of Cambridge, Dissertation, Kings College
Karger CR (1996) Wahrnehmung und Bewertung von "Umweltrisiken". Was können wir aus Naturkatastrophen lernen? Arbeiten zur Risiko-Kommunikation. Heft 57. Programmgruppe Mensch, Umwelt, Technik, Forschungszentrum Jülich GmbH, Jülich: März 1996
Klingmüller O, Bourgund U (1992) Sicherheit und Risiko im Konstruktiven Ingenieurbau. Vieweg, Wiesbaden
NYSDOT (2004) Bridge safety assurance: hydraulic vulnerability manual. Department of Transportation, New York State
Proske D (2009) Catalogue of risk. Springer, Berlin
Riedl R (2000) Strukturen der Komplexität – Eine Morphologie des Erkennens und Erklärens. Verlag, Heidelberg
Taricska MR (2014) An analysis of recent bridge failures in the United States (2000–2012). MSc. thesis, The Ohio State University
Vogel T, Zwicky D, Joray D, Diggelmann M, Hoj NP (2009) Tragsicherheit der bestehenden Kunstbauten, Sicherheit des Verkehrssystems Strasse und dessen Kunstbauten, Bundesamt für Strassen, Dezember 2009, Bern
Wardhana K, Hadipriono FC (2003) Analysis of recent bridge failures in the United States. J Perform Constr Facil ASCE, 144–150
Weinberg GM (1975) An introduction to general systems thinking. Wiley-Interscience, USA
Wolhuter KM (2015) Geometric design of roads handbook. CRP Press, Taylor & Francis Group, Boca Raton

Chapter 3
Method

In this book mainly the technique of comparison is used in terms of a meta-analysis. Meta-analyses are common in other fields of science such as medicine, however they are rarely used in engineering sciences. Meta-analyses are usually applied if own samples are simply too expensive, too time-consuming or unavailable due to other restrictions such as ethical, social or technical limitation.

In general, a meta-analysis is a technique to combine results of former studies. The meta-analysis should provide an extension of the individual investigated populations (more samples, spatial and temporal extension) and usually an extension of the parameter space which has been investigated (local and global construction materials etc.). This approach should yield to more robust results and should resolve the limitations of the individual studies. Furthermore the extension of the parameters, such as a larger time scale, may yield to other conclusions compared to individual studies using only short time scales.

In our case the extension of the population and parameters may consider other time scales (years, decades, centuries), different regions or countries (U.K., U.S., Switzerland, Germany etc.), different building materials, different construction techniques, different natural hazards which have a return period of decades or centuries, different maintenance techniques and others.

D. Proske, *Bridge Collapse Frequencies versus Failure Probabilities*, Risk Engineering,
https://doi.org/10.1007/978-3-319-73833-8_3

Chapter 4
Categorization of Bridges

4.1 Introduction

The systematic analysis of the background collapse data can provide detailed information about the specific type of bridge structures or bridge material contributing to the overall collapse data over-proportionally or under-proportionally. This conclusion would provide the way to further identify weaknesses in the design, construction and maintenance of the bridges and to resolve these weaknesses. However, to carry out these steps the bridge population has to be subdivided into different sub-populations. The parameters for the subdivision are introduced in this chapter.

4.2 Structural Systems of Bridges

The ways to design bridges in terms of statical systems are limited. Schlaich (2003) has introduced a typology of bridges as shown in Fig. 4.1. However the various possible solutions have been used with different frequencies. Some solutions such as simple beam structures are very frequently used whereas suspension bridges are used on a smaller number. Figures 4.2 and 4.3 show the distribution of the structural types related to the overall bridge surface area on U.S. highway bridges and to the overall bridge number in the U.S.

Historically, the statical systems were determined by the mechanical properties of the building material, usually taken from the vicinity, and by the construction technology available. Of course both the material and the construction technology are related.

However, with the development of new building materials such as steel and composite materials such as reinforced concrete, the possibilities for bridge design increased and traditional limits were exceeded. This is not only true by the sheer dimensions and spans (see Fig. 4.4), but also by a shift from structures mainly exposed

D. Proske, *Bridge Collapse Frequencies versus Failure Probabilities*, Risk Engineering,
https://doi.org/10.1007/978-3-319-73833-8_4

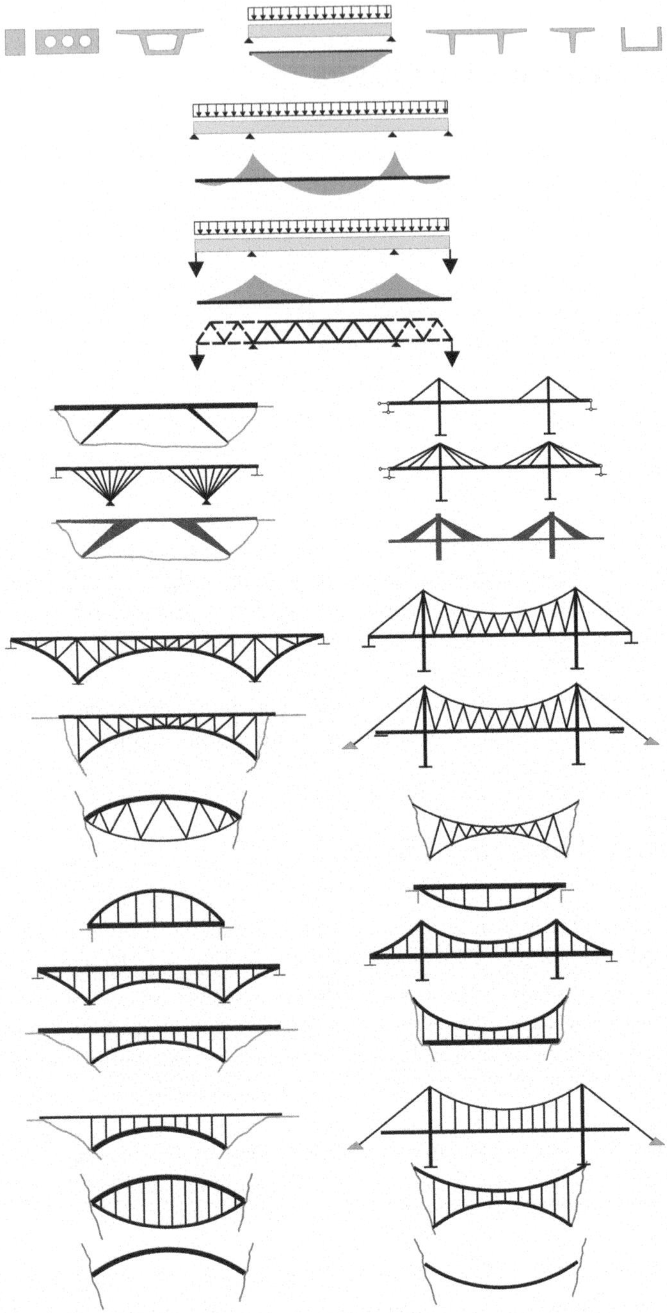

Fig. 4.1 Typology of bridges according to Schlaich (2003)

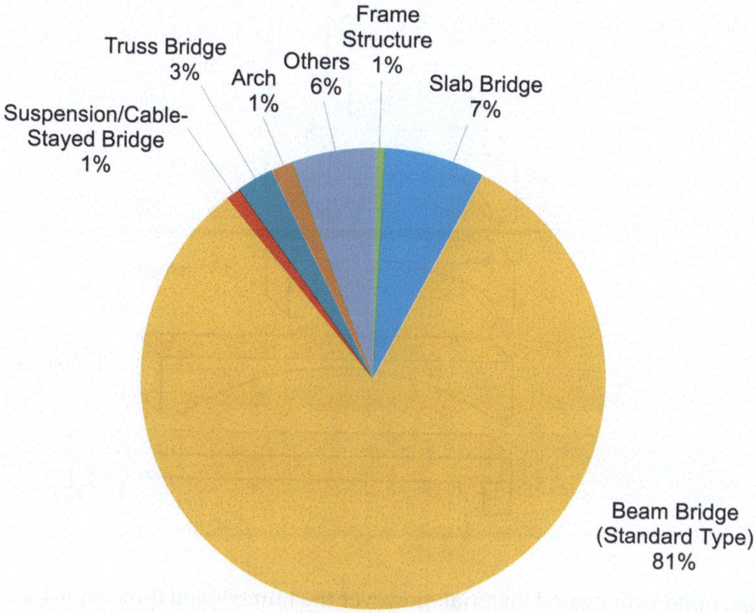

Fig. 4.2 Percentage of bridge deck area related to the structural type of the bridges on U.S. highways (Steinbock et al. 2016)

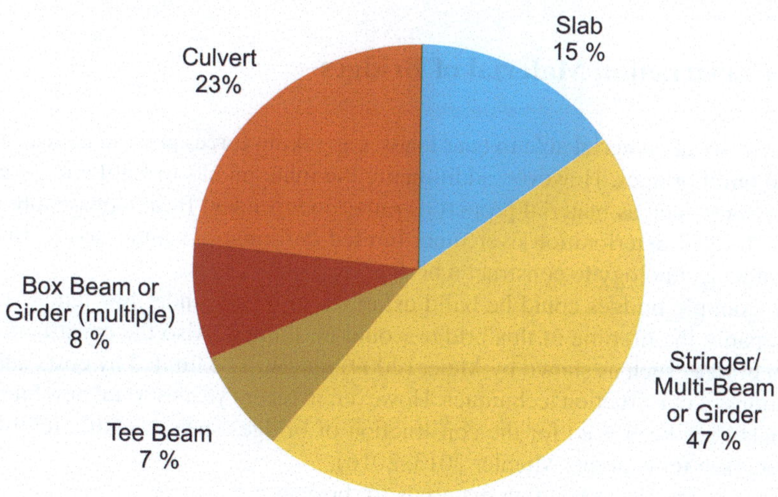

Fig. 4.3 Percentage of bridge numbers related to the structural type of the bridges in the U.S. (Hersi 2009)

to normal forces such as arch bridges and suspension bridges to structures mainly exposed to bending moments such as all types of beams. Of course, beam structures

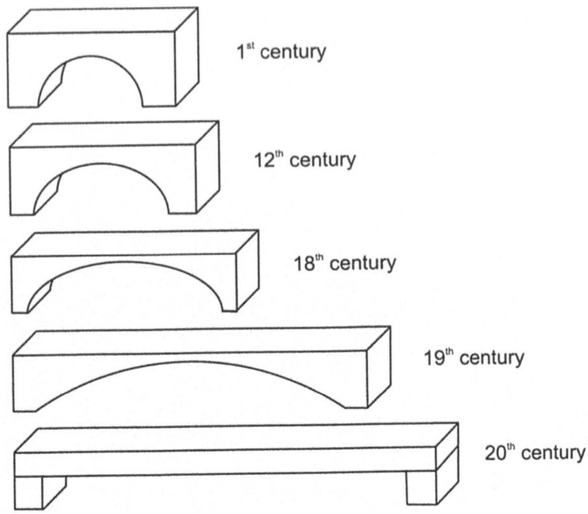

Fig. 4.4 Development of massive bridge shapes over time (Proske and van Gelder 2009)

were also build using wood material, however the number and dimensions are rather limited compared to current standards. The shift is not only related to the new construction material and highly sophisticated construction technologies but also by the impact of the construction costs on the overall costs.

4.3 Construction Material of Bridges

Theoretically all material able to bear loads, e.g. taking forces and moments, can be used to build bridges. However, additionally the material has to fulfil some further requirements such as material properties rather independent from temperature conditions, limited deterioration over time, limited deformations under loads, limited costs and a technology to construct a bridge.

For example bridges could be build using ice in winter under low temperatures but probably the lifetime of this bridge would be limited. Also the construction of carbon bridges such as shown by Meier (2009) may still be limited by costs and the limitations of construction techniques. However, in recent years several new building materials have been used for the construction of bridges such as textile reinforced concrete (see for example Michler 2013, 2016).

The most important building materials for bridges are:

- Wood
- Steel and Composite Material
- Concrete and Reinforced Concrete
- Pre-stressed Concrete
- Stone and Masonry
- Carbone Fibre and Plastics.

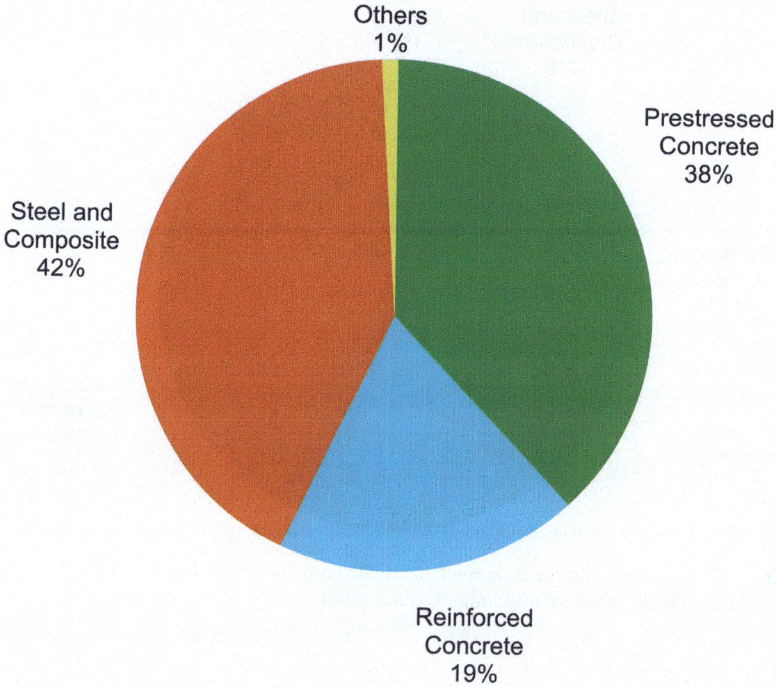

Fig. 4.5 Percentage of bridge deck area related to the construction material for bridges on U.S. highways (Steinbock et al. 2016)

Concrete is by far the most popular construction material worldwide. It has been estimated that about 1 m^3 concrete is produced worldwide for every single human per year.

Figure 4.5 shows the distribution of the construction material related to the bridge deck area for U.S. highway bridges and Fig. 4.6 shows the same for German highway bridges.

Neither ratio is constant over time. Figure 4.7 shows the ratio of the design-type railway bridges over time for the German railway excluding wooden railway bridges. Wooden railway bridges were not permitted after 1865 based on technical requirements of the association of German railway administrations. Wooden railway bridges were built until 1860 as permanent structures with a span up to 40 m. In many other countries, wooden structures made a significant part of the bridge stock (Weber 1999). Figures 4.8 and 4.9 evidently show the change of the ratio of the construction materials by giving the distribution of the construction material for the overall bridge stock of the German railway and for all bridges newly constructed since 1991. We can conclude, in Germany concrete is by far the most frequent used construction

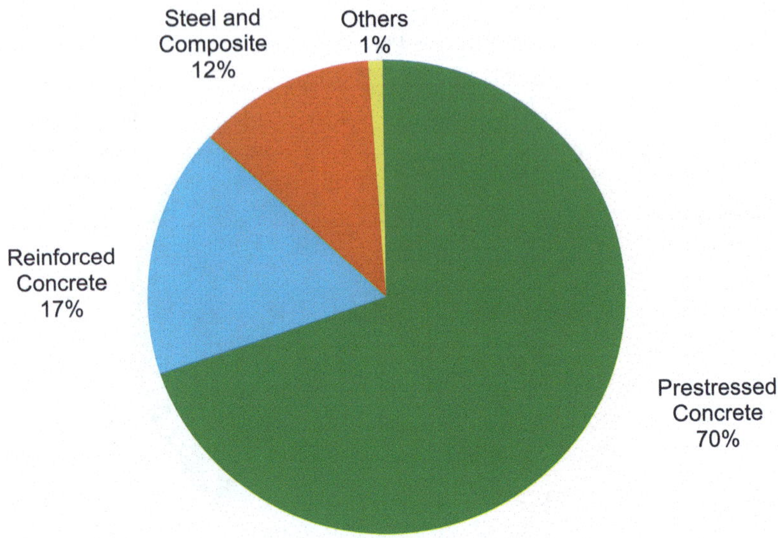

Fig. 4.6 Percentage of bridge deck area related to the construction material for bridges on German highways and federal state roads (Steinbock et al. 2016)

material for bridge constructions for both, highway and railway bridges. In the U.S., steel contributes more than in Germany but still concrete provides more than 50%.

4.4 Construction Method of Bridges

There exists a variety of construction techniques related to construction materials, for example for concrete bridges. Concrete bridges can be constructed precast, in-situ or mixed. Every one of theses techniques includes further sub-techniques, such as cast in-situ post tensioned, balanced cantilever, incrementally launched and further techniques for in-situ concrete construction, see the relevant publications for example Stritzke (2007), Wittfoht (1980) and Trayner (2007).

Detailed information would be helpful if bridge collapses are dominated by the construction methods. However, no or only limited data is available, for example see Xu et al. (2016). Therefore no further details are included here.

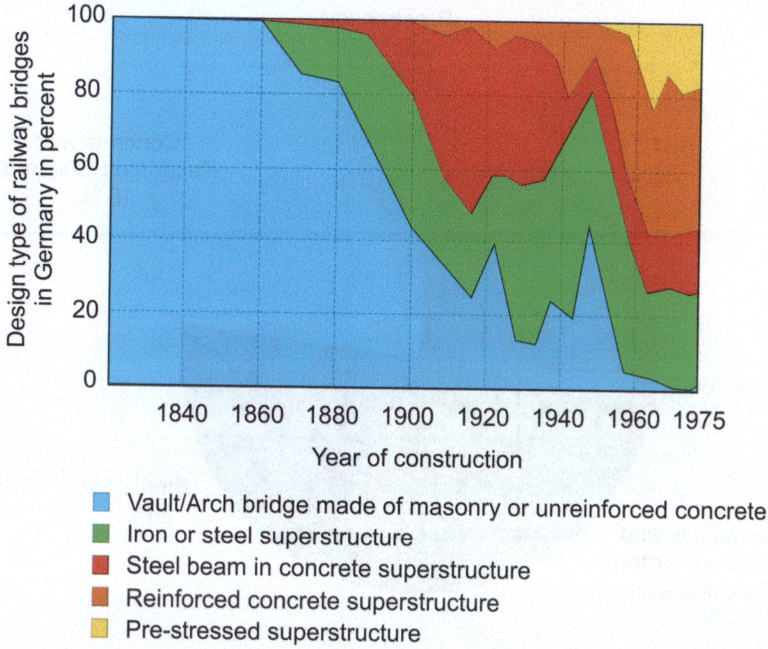

Fig. 4.7 Design-type railway bridges in Germany according to Weber (1999) excluding wooden bridge structures

4.5 Age Distribution of Bridges

The age of bridges may be a potential early indicator for the collapse of bridges. There exist various publications regarding the age distribution of bridges in industrialised countries because this information is required for maintenance management. Figure 4.10 gives the age distribution for the U.S. bridge stock over the last decades. Figure 4.11 gives the age distribution for the German highways, whereas Fig. 4.12 gives the age of bridges for the Germany city Düsseldorf as example for municipal bridge stock. Figure 4.13 gives the age distribution for the German railway bridges. Marx (2009) states, that the oldest bridge which is owned by the Germany railway was constructed in 1779.

Figure 4.14 brings together all data available. The age of bridges in Germany clearly reflects major political events, such as World War II or the German reunification. The German railway data also indicates the strong economic growth at the end of the 19th century in Germany and the increase in mass transport around World War I. In contrast, the U.S. history was more stable over the last 200 years, except for the American Civil War between 1861 and 1865. Therefore it is concluded that no such strong fluctuations of the age distribution have to be assumed.

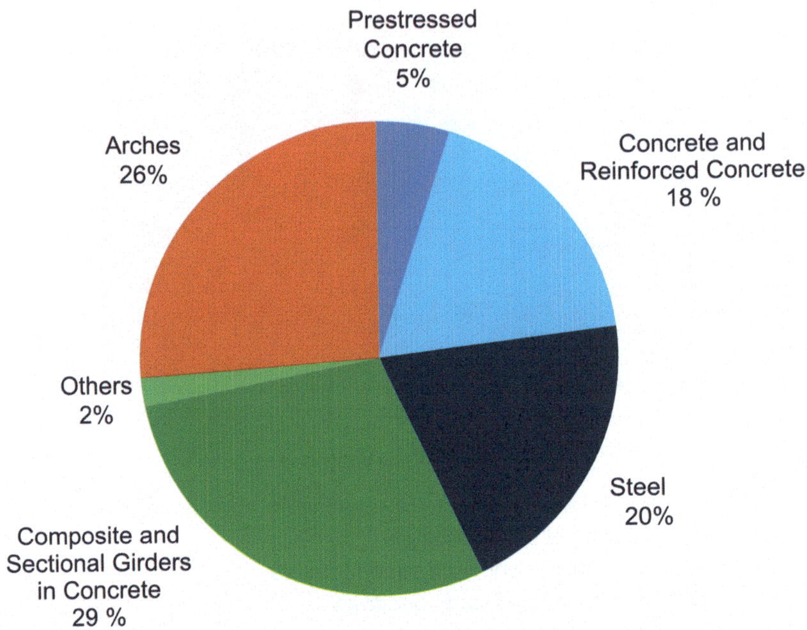

Fig. 4.8 Percentage of German railway bridges related to the construction material (Garber 2009)

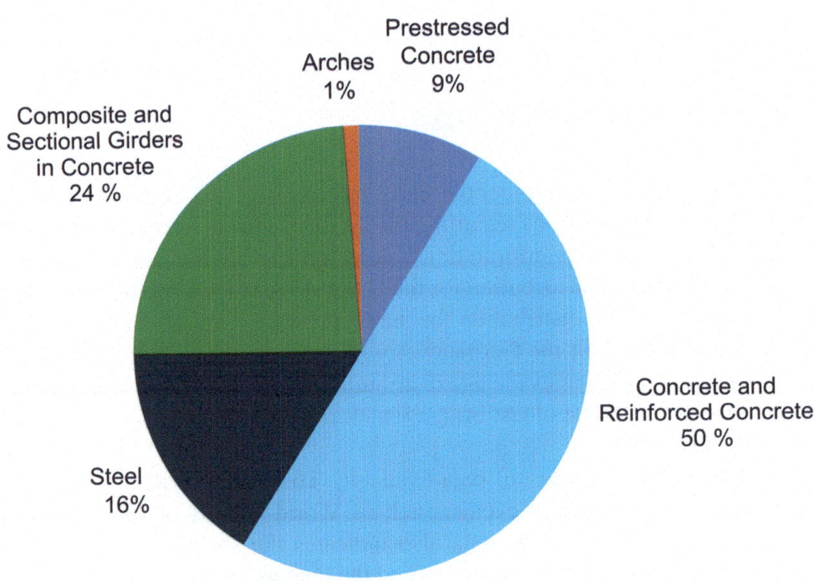

Fig. 4.9 Percentage of German railway bridges related to the construction material built since 1991 (Marx 2009)

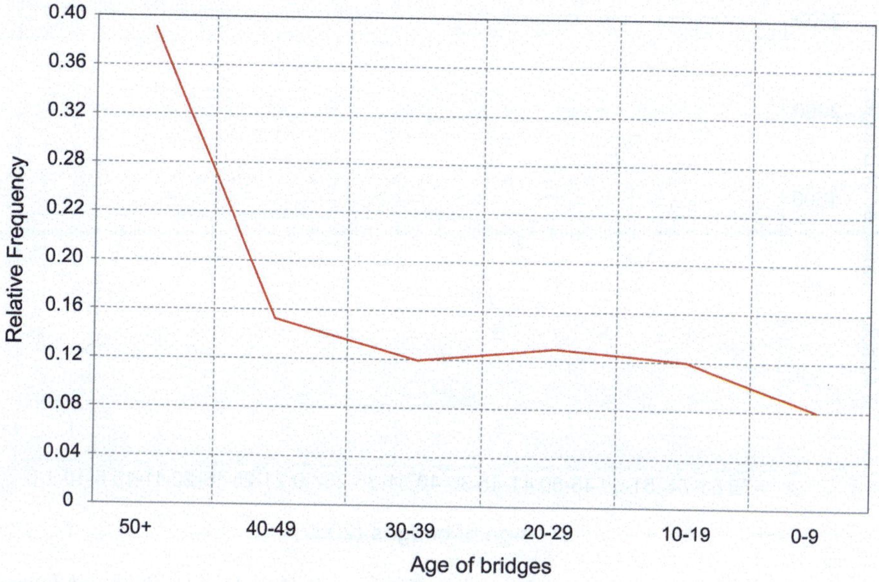

Fig. 4.10 Age distribution of bridges in the U.S. (ASCE 2017)

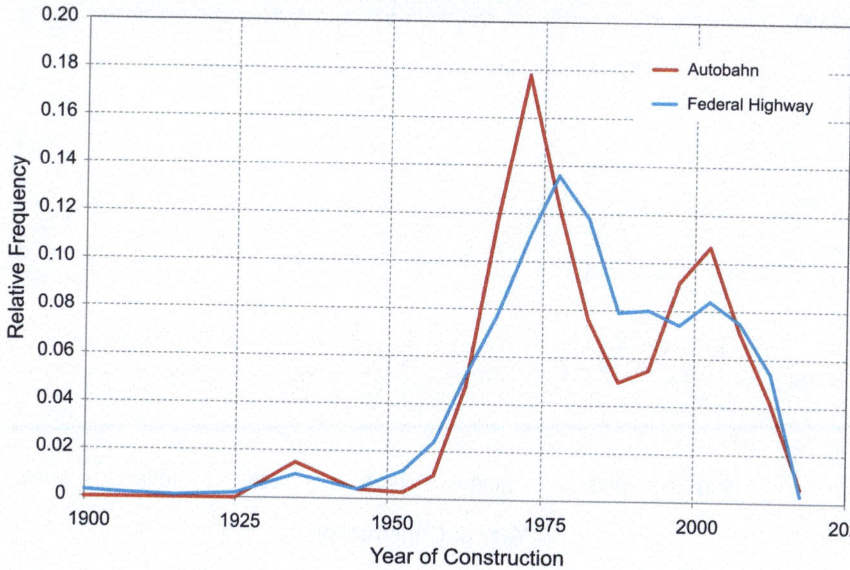

Fig. 4.11 Year of bridge construction of German highway bridges (BAST 2017)

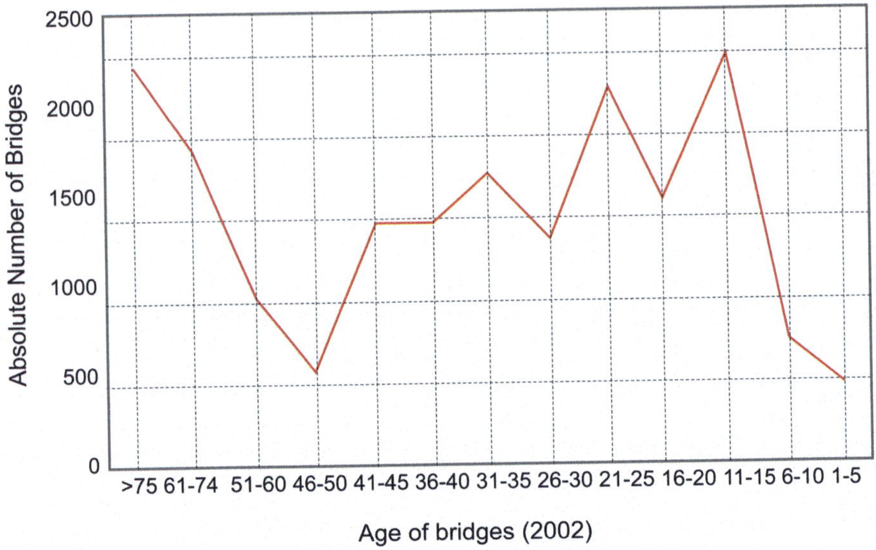

Fig. 4.12 Age distribution of road bridges in the German city Düsseldorf (Vollrath and Tathoff 2002)

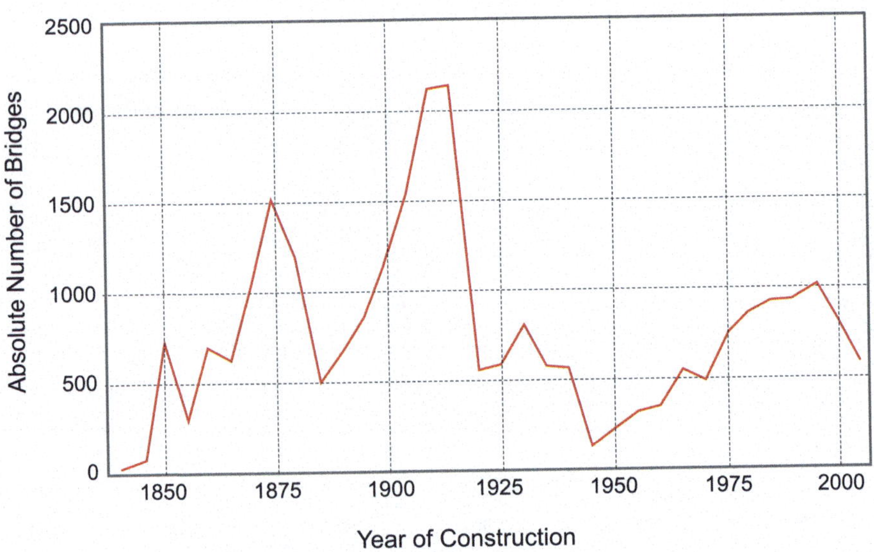

Fig. 4.13 Year of bridge construction of German railway bridges (Garber 2009)

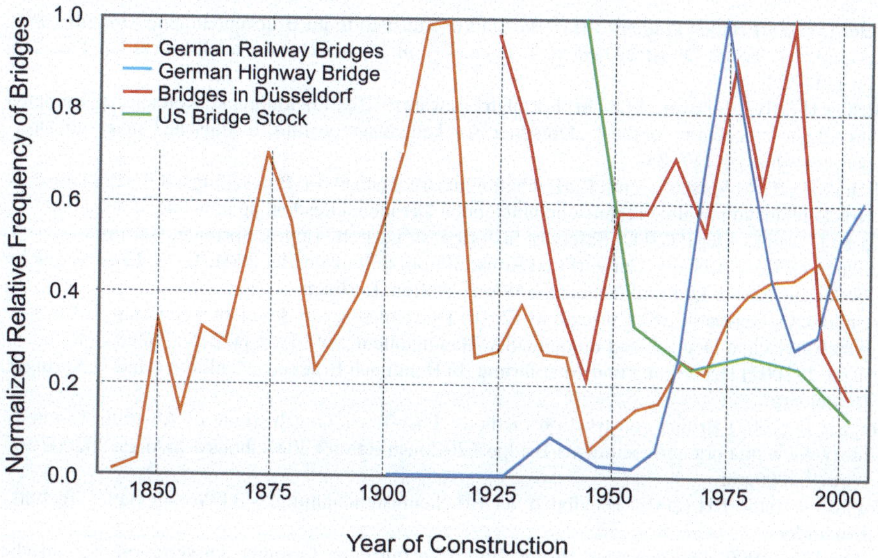

Fig. 4.14 Year of bridge construction based on Figs. 4.10, 4.11, 4.12 and 4.13 (The steep drop in the curves for Düsseldorf and the U. S. bridge stock in the left-hand area is due to cumulative values)

4.6 Conclusion

The subdivision of the bridge population is mainly done considering the structural type of the bridges, the bridge material and the construction method. However the construction method itself should mainly be related to collapses during the construction, otherwise a double counting may exist, since certain structural systems may be related to certain construction techniques.

References

ASCE (2017): 2017 infrastructure report card: bridges: overview. American Society of Civil Engineers, http://www.infrastructurereportcard.org

BAST (2017) Brücken an Bundesfernstrassen, http://www.bast.de/DE/Statistik/Bruecken/Brueckenstatistik.pdf?__blob=publicationFile&v=7

Garber S (2009) Brücken für Europa, In: Eisenbahnbrücken - Ingenieurbaukunst und Baukultur. In: Mehdorn H, Schwinn KH (eds) DVV Media Group, Eurailpress, Hamburg, pp 67–71

Hersi M (2009) Analysis of bridge failure in United States (2000–2008). Ohio State University, Master of Science Thesis

Marx S (2009) Die Spezifik der Eisenbahnbrücken, In: Eisenbahnbrücken - Ingenieurbaukunst und Baukultur, Eds. H Mehdorn & KH Schwinn, DVV Media Group, Eurailpress: Hamburg, pp 51–57

Meier U (2009) Haben kohlenstofffaserverstärkte Kunststoffe im Brückenbau eine Zukunft?, Proceedings of the 19. Dresdner Brückenbausymposium, Technische Universität Dresden, Dresden, pp 191–208
Michler H (2013) Innovativ! Leicht! Formbar! Bewährt! - Textilbetonbrücke Rottachsteg Kempten im Allgäu. Proceedings of the 23. Dresdner Brückenbausymposium, Technische Universität Dresden, Dresden, pp 239–252
Michler H (2016) Verstärken mit Carbonbeton im Brückenbau. In: Proceedings of the 26. Dresdner Brückenbausymposium, Technische Universität Dresden, Dresden, pp 235–248
Proske D, van Gelder P (2009) Safety of historical stone arch bridges. Springer, Berlin
Schlaich J (2003) Brücken: Entwurf und Konstruktion, Betonkalender 2004, Teil 1, Ernst und Sohn Verlag für Architektur und technische Wissenshaften, Berlin, pp 1–26
Steinbock O, Garibaldi MP, Curbach M (2016) Der Umgang mit dem Brückenbestand - Ein Vergleich zwischen Deutschland un den USA, Bauingenieur, vol 91(6), pp 215–226
Stritzke J (2007) Herstellung und Ausführung. In Handbuch Brücken, Mehlhorn (Hrsg.), Springer, Heidelberg
Trayner D (2007) Bridge construction methods. Post-Tensioning Institute of Australia Limited. http://www.ptia.org.au/Documents/Bridge%20Construction%20Methods%20Aug%202007%20rev%2000.pdf
Vollrath F, Tathoff H (2002) Handbuch der Brückeninstandhaltung, VBT Verlag Bau + Technik: Düsseldorf
Weber WL (1999) Die gewölbte Eisenbahnbrücke mit einer Öffnung. Dissertation, Technische Universität München, München
Wittfoht H (1980) Bridges: construction techniques and methods (concrete and steel). In: IABSE proceedings, vol 4(P-29), pp 49–64
Xu FY, Zhang MJ, Wang L, Zhang JR (2016) Recent highway bridge collapses in China: review and discussion. J Perform Constr Facil 30(5):8 p

Chapter 5
Measures of Safety

5.1 Introduction

Bridges form a substantial part of the infrastructure systems in almost all countries worldwide. The global stock is estimated between five and six million bridges (see Chap. 6). Approximately one bridge per 500 inhabitants is counted in developed countries and about one bridge per 2000 inhabitants is counted in developing countries.

Bridges are designed to function safely over an extreme long period, usually 100 years. However many cases are known in which bridges function more than hundred years, in some cases more than a thousand years.

Whereas early design concepts of bridges were purely based on empirical rules, modern safety concepts of bridges are based on certain reliability or risk measures (Proske and van Gelder 2009). One of the most important reliability measures are the computed probability of failure and the safety index β, which is a substitute of the probability of failure (see Table 5.1). For new as well as existing bridges target values of the probability of failure and the safety index respectively are given in certain codes and guidelines.

Besides the calculated probability of failure, the observed frequency of bridge collapse can be calculated based on the number of bridge collapses and the overall bridge stock for a certain period and region. Whereas the probability of failure is a bottom-up approach, the collapse frequency is a top-down approach. Theoretically both parameters should yield to the same result. This chapter deals with the calculation of the probability of failure.

D. Proske, *Bridge Collapse Frequencies versus Failure Probabilities*, Risk Engineering,
https://doi.org/10.1007/978-3-319-73833-8_5

Table 5.1 Conversation between probability of failure and safety index

P_f	10^{-11}	10^{-10}	10^{-9}	10^{-8}	10^{-7}	10^{-6}	10^{-5}	10^{-4}	10^{-3}	10^{-2}	10^{-1}	0.5
β	6.71	6.36	5.99	5.61	5.19	4.75	4.26	3.72	3.09	2.33	1.28	0.0

5.2 Probability of Failure

5.2.1 Unconditional Probability of Failure

First proposals about probabilistic based safety concepts can be found by Mayer (1926) in Germany and Chocialov in 1929 in the Soviet Union (Murzewski 1974). In the thirties of the 20th century already the number of people working in that field had increased, just to mention Streleckij in 1935 in the Soviet Union, Wierzbicki in 1936 in Poland and Prot in 1936 in France (Murzewski 1974). Already in 1944 in the Soviet Union the introduction of the probabilistic safety concept for structures was forced by politicians (Tichý 1976). The development of probabilistic safety concepts in general experienced a strong impulse during and after the World War II, not only in the field of structures, but also in the fields of aeronautics (Pugsley 1968). Freudenthal (1947) published in 1947 his famous work about the safety of structures. Since then, probabilistic safety concepts have undergone considerable further development. Even a model code for the probabilistic safety concept of structures has been published by the JCSS (2004).

The probability of failure as proof measure for safety can be referred to one year or the life time of the structures, the equations have already been introduced in Chap. 2.

The probability of failure is usually related to a limit state equation, in other terms it is related to one single proof equation during the various proofs required to confirm the safety of a structure as a whole. It considers all random variables of the resistance and the loading in the investigated proof equation.

Since more than a decade, there exist several commercial as well as research based software tools to compute the probabilities of failure of structures. The following list gives the names of some software programs and the references define the state of the software more than ten years ago:

- UNIPASS (Lin and Khalessi 2006),
- ProFES (Wu et al. 2006),
- Proban (Tvedt 2006),
- PHIMECA (Lemaire and Pendola 2006),
- PERMAS-RA/STRUREL (Gollwitzer et al. 2006),
- NESSUS (Thacker et al. 2006),
- COSSAN (Schueller and Pradlwarter 2006),
- CalRel/FERUM/OpenSees (Der Kiureghian et al. 2006),
- ANSYS PDS und DesignXplorer (Reh et al. 2006),
- ATENA/SARA/FREET (Pukl et al. 2006),
- VaP (Petschacher 1994),

- OptiSlang (Schlegel and Will 2007),
- PERMAS (Intes GmbH, Stuttgart),
- ISPUD (University of Innsbruck, Innsbruck),
- RACKV (University of Natural Resources and Applied Life Sciences, Vienna).

Since then, software has experienced further developments to improve the handling of the software as well as the computation capabilities. Nowadays large non-linear finite elements models can be included into probabilistic computations giving probabilities of failure in a reasonable time. Some effort was undertaken to develop a generalized approach for probabilistic software, see Epstein et al. (2008).

Many of the above listed programs can be downloaded free of charge for test runs. The software is in many cases validated on test samples.

Even without commercial programs, simple First Order Reliability Method (FORM) computations can be carried out with standard spreadsheet software including an optimisation tool like the solver in EXCEL. With this tool it is possible to compute the safety index. An example of such an application can be found in Low and Teh (2000). The most spreadsheets also provide random number simulations and inverse probability distribution functions and therefore allow Monte-Carlo-Simulations.

It can be concluded that the computation of the unconditional probability of failure is state-of-the-art and that the required tools are available.

5.2.2 Conditional Probability of Failure

In contrast to the unconditional probability of failure, the conditional probability of failure does not consider all input variables as random input variables. Usually the loading is excluded from the random variable vector. This probability is often called fragility. Fragilities are functions of the probability of failure with regard to intensity measures of a load, often a seismic loading in terms of a spectral acceleration (Fig. 5.1) or a flood loading in terms of a water gauge. In some industries fragilities are very popular since they can be easily implemented in Fault Trees, for further details and examples see EPRI (1994), Kennedy et al. (1980), Kennedy (1999), Zentner et al. (2008), Proske (2012).

The general fragility function is defined as:

$$P_f(a) = A_m \times \varepsilon_R \times \varepsilon_U$$

with ε_R as aleatoric uncertainty and ε_U as epistemic uncertainty. Considering the fragility function type as lognormal distribution, which is a very common approach, it yields to:

$$P_f(a) = \Phi\left[\frac{\left(\log\left(\frac{a}{A_m}\right) + \beta_U \cdot \Phi(Q)\right)}{\beta_R}\right]$$

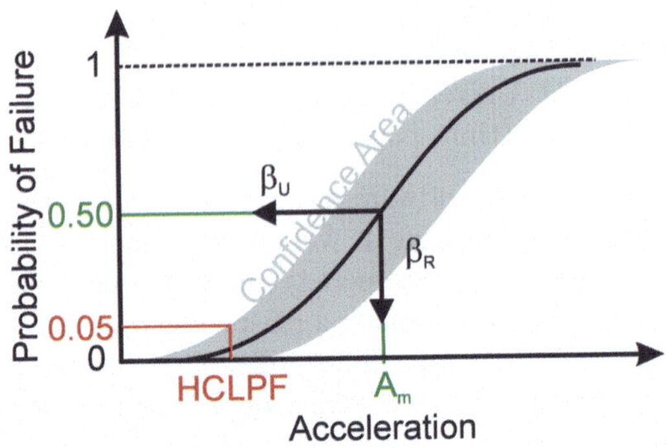

Fig. 5.1 Fragility function $P_f(a)$ (probability of failure for a spectral acceleration), for further terms see text

with Φ as standard normal distribution, A_m as median intensity of the loading, a as intensity value of the loading, Q as confidence interval and β_u and β_R as uncertainty measures.

Lower fractile values of the fragility functions are called *HCLPF*-values (high confidence of low probability of failure) and are comparable to characteristic values in structural engineering.

The *HCLPF*-value is computed as

$$HCLPF = A_m \cdot \exp(-1.645 \cdot (\beta_R + \beta_U))$$

or

$$HCLPF = A_m \cdot \exp(-2.3 \cdot \beta_C)$$
$$\beta_C = \sqrt{\beta_R^2 + \beta_U^2}.$$

Examples for the computation of seismic fragilities for bridges can be found in Banerjee and Shinozuka (2008), Pan et al. (2007) and Billah and Alam (2013).

5.3 Risk Measures

5.3.1 Introduction

The parameter "probabilities of failure" belongs to the so-called risk measures or risk parameters of zero order. Usually risk is defined as the product of the probability or frequency of an event and the amount of damage of this event. Some authors also take into account scenarios but we leave that aside for simplification. By comparing the definition of risk and the definition of the probability of failure it can be clearly seen that the latter does not include a damage evaluation, the probability of failure simply states that the target value or threshold respectively is exceeded.

In the following sections several risk parameters are introduced very briefly before the target values of probabilities of failure are discussed in detail. The chapter shows that useful and adequate risk parameters have been developed and that based on these risk parameters bridges are a safe technical product.

5.3.2 *Mortality*

Mortality is one of the most easy to understand risk parameters. The parameter describes the ratio of the number of fatalities related to a certain event and to the overall population. Mortalities are used in many fields, such as medicine, statistics etc. Some values also refer to bridge collapses and structural collapses.

Blockley (1980) gives an annual mortality risk of 10^{-7} per year for the public related to structural collapse including buildings. This value is further referenced and used in Haldi and Vulliet (1998), Imhof (2004), Proske (2009) and Vogel et al. (2009). Reid (2000) gives a value of 1.4×10^{-7} per year for structural collapse mortalities.

Tsang and Wenzel (2016) and Tanner and Hingorani (2015) and Hingorani (2017) give acceptable mortalities for structures under specific loads (gas explosion).

Das (1997) gives a mortality risk for the total U.K. population of 0.2×10^{-8} per year for bridge collapse. Vogel et al. (2009) estimate the mortality risk with 1×10^{-8} per year for bridge collapse.

O'Connor and Shaw (2000) give an acceptable annual risk of accidental death due to structural bridge failure of 10^{-6} per year. Menzies (1996) gives an acceptable risk of loss of life caused by bridge collapse in the range of 10^{-6} per year. Both values represent acceptable mortality values.

Tanner and Hingorani (2015) state that consensus of opinion is that the individual risk to persons due to structural collapse should be limited to values ranging from 10^{-6} to 10^{-5} per year. They refer to a former version of the ISO 2394 (1998) giving an acceptable individual risk due to structural collapse of 10^{-6} per year, to the Dutch Ministry of Housing, Spatial Planning and the Environment giving 10^{-6} per year for new and 10^{-5} per year for existing structures (Vrijling et al. 2005). These values are also the basis for the development of target reliability levels for existing structures (Steenbergen and Vrouwenvelder 2010; Sýkora et al. 2013).

5.3.3 *Fatal Accident Rate*

The parameter Fatal Accident Rate (FAR) relates the mortality to the exposure time. This is in contrast to the sheer mortality value which is usually a number related to a calendar year. Therefore the mortality parameter is true for all risks to which we are

Table 5.2 FAR Target values based on various references

Reference	FAR target value	Remark
Randsaeter (2000)	15	Oil industry
Aven et al. (2005)	10	Installation based on the NORSOK Z-103 code
Cox et al. (1990)	4	British industry
Cox et al. (1990)	0.4	Any particular hazard
Mannan (2005)	3.5	Industry
Menzies (1996)	2.0	Related to bridge collapse
Mannan (2005)	0.35	Any particular hazard
Maag (2004)	0.04…0.12	Fire risk in buildings for the public (Switzerland)
Maag (2004)	0.05…0.3	Fire risk in buildings for the public (Norway)

more or less exposed over the entire year, for example health risks. We even may be exposed to risk of structural failure most time of the year, but not to the collapse of bridges. The exposure time of building collapse is approximately more than 20 h per day; the exposure time of bridge collapse is usually not more than a few minutes per day for most people. There is almost a factor of two orders of magnitude difference between bridge collapse exposure time and building collapse exposure time. Please keep in mind, the target probability of failure is more or less equal for both types of structures (see Sect. 5.4).

The Fatal Accident Rate relates the mortality value to an exposure time of 10^8 hours (Proske 2009). Values of the FAR are in the range 2000 for Alpine climbing to 0.0002 killed by an airplane falling down. The FAR for building collapse is in the range of 0.0020 (Proske 2009; Menzies 1996). Splitting this FAR value into the ratio of exposure time between buildings and bridges, the FAR for bridges would be in the range of 0.00002, which is extremely low.

Table 5.2 lists several FAR target values. There are target values for industries such as the oil industry and target values for the public. In general, the target values for the public are at least one order of magnitude lower than the target values for industries. However even the target values for the public show itself a variety of nearly one order of magnitude—see the values given in Maag (2004).

Keeping in mind the very low value of FAR for bridge collapse in the range of 0.00002, this value by far fulfils the target values, whether they are 0.04 or 0.4 for the public.

Das (1997) gives a FAR of 0.1 for the travelling population of the U.K. based on 25×10^6 people travelling on average one hour a day. Travelling parameters (for Germany) are shown in Table 5.3. It can be seen that most parts of the population are travelling which would yield a number of more than 70×10^6 people for the travelling population in Germany. Assuming a slightly higher travelling participation in Germany, the FAR would still be comparable to the value from the U.K. Comparing

Table 5.3 Development of mobility parameters for Germany (Chlond et al. 1998)

Indicator	1976	1982	1989	1992	1994	1995	1996	1997
Traffic participation in % of population	90.0	82.2	85.0		91.9	93.9	92.9	92.0
Number of paths per person per day	3.09	3.04	2.75	3.13	3.32	3.39	3.46	3.52
Number of paths per mobile person per day	3.43	3.70	3.24		3.61	3.61	3.73	3.82
Passenger cars per inhabitant				0.508	0.502	0.467	0.511	0.518
Travelling time per day in hours: minutes	1:08	1:12	1:01		1:19	1:20	1:21	1:22
Kilometre travelled per person and day	26.9	30.5	26.9	33.8	39.3	39.2	39.6	40.4
Average distance travelled per path in kilometre	8.7	10.0	9.80	10.8	11.8	11.5	11.5	11.5

the FAR of 0.1 with the 0.00002 for bridge failure indicates a very low number for bridge failure compared to other travelling related risks.

More details about FAR values including various examples can be found in Proske (2009).

5.3.4 *F-N-Diagrams*

Both, mortalities and Fatal Accident Rates do not consider the severity of a single collapse. However, it is well known that the public considers one large accident with a certain number of fatalities more severe than many single smaller accidents with the same overall number of fatalities. This is called risk aversion and can be considered in so-called Frequency-Number-diagrams or F-N-diagrams. A detailed introduction can be found in Proske (2009) and elsewhere (Ball and Floyd 2001, Proske 2009).

There exists a great variety of target curves for such diagrams as shown in Ball and Floyd (2001, Proske 2009). Figure 5.2 shows an example.

In Proske (2009) the author has computed the F-N curve for two individual bridges exposed to ship impact based on worldwide ship accident data (Fig. 5.3). These two curves are then compared to the variety of target curves. It can be seen that the curves are in the upper part of the variety of target curves, which is also the ALARP (as low as reasonably practicable) region. Therefore in these specific case, bridge collapse prevention measures had to be applied and indeed, were implemented.

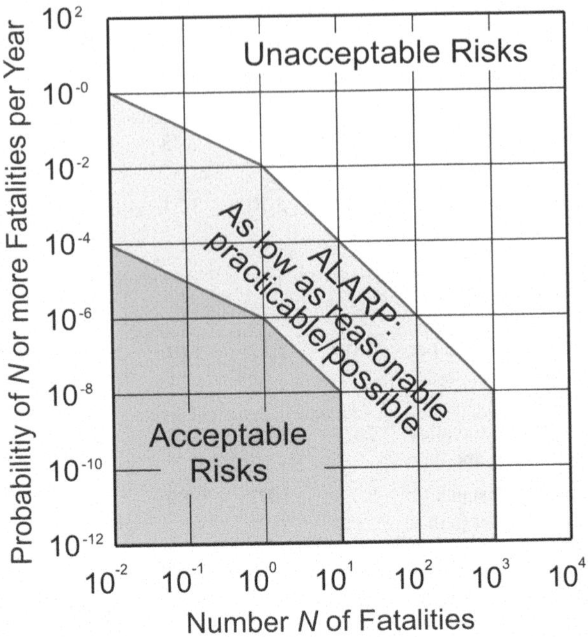

Fig. 5.2 Example of target lines in a F-N-diagram with three regions

F-N-diagrams have been frequently used for risk evaluation of dams (Jonkman et al. 2003; Jonkman 2007; ICOLD 2005) and recently for the risk evaluation of structures under seismic loading (Daniel et al. 2017; Tsang and Wenzel 2016). Tanner and Hingorani (2015) refer to F-N-diagrams for buildings as given by Vrijling et al. (2005). Tanner and Hingorani (2015) also refer to ISO 2394 (1998). No other application of F-N-diagrams for bridge collapse is known besides Fig. 5.3.

5.3.5 *Lost Life Years*

The risk parameter Lost Life Years (LLY) compares the age of humans killed during an event to the average life expectancy. Therefore this parameter can consider whether rather young people are killed or older ones. Furthermore the parameter can consider injuries and disabilities within the so-called Disability Adjusted Life Years (Proske 2009).

To calculate the LLY for bridges, information about the age population and the ratio of injured to killed people are required.

The ratio of injured to killed people during a structural collapse is about six (Rackwitz 1998). For comparison, the ratio of injured to killed is about 20 for car accidents and significantly smaller than one for airplane crashes.

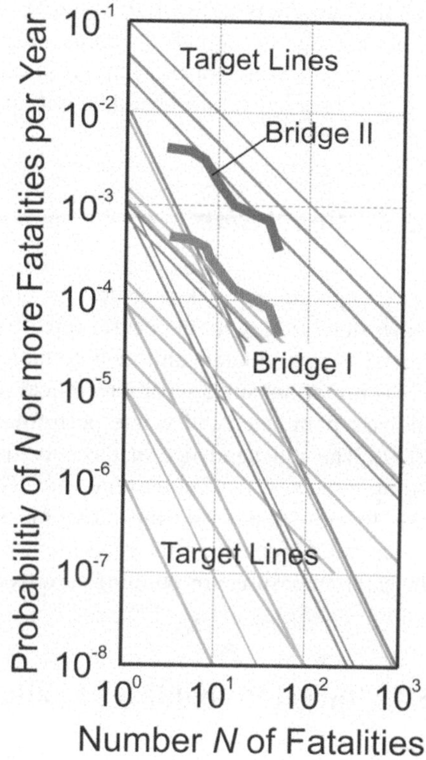

Fig. 5.3 Computed *F*-*N*-curves for two bridges under ship impact including various target lines

If one assumes that this ratio is the same for building collapses and bridge collapses, it can be concluded that the Disability Adjusted Life Years will probably be more than twice the Lost Life Years due to the killed people (assuming a disability of 30% after the collapse and the ratio of fatalities to injured people one receives $1{:}6 \times 0.3 = 1.8$).

Furthermore the age distribution of people crossing bridges is probably slightly higher for the age cohorts between 20 and 60 years since people working are probably more travelling than others.

It is further assumed that the changes of the traffic volume can be directly related to the exposure time to bridge collapse. The traffic volume changes over the day, over weekdays, over month and it is related to economic cycles. Therefore the number of bridge crossings is time-dependent (Krystek and Zukowska 2005; SBA 2006). Additionally, there seems to be a trend of general grows of traffic (see Table 5.3).

Considering these facts, the average age of the fatalities killed during a bridge collapse and the average age of the injured are probably close to the average age of the population; the absolute number of killed people due to bridge collapse is very low (see Sect. 5.3.2). Therefore the parameter of Lost Life Years will not show a significant risk increase. Whereas some social risks indicate several years lost, some technical products show up to hundreds of days lost, bridge collapse will probably be in the range of seconds or minutes lost.

Target values are not known for bridge collapses but it seems reasonable to assume that the observed values would be within the range of the target values.

5.3.6 Conclusion

The computation of risk measures for bridges and the comparison with target values shows that bridges are in general safe products. This conclusion is true for all investigated risk parameters although no target value is known for Lost Life Years. This fact is important since several technical products show excellent performance under one risk parameter and worse performance under another risk parameter (Proske 2009). The independence of the conclusion from the selected risk parameter is a strong indicator, that bridges are safe. Therefore in the following section we rather have to show that our design safety measure functions, not that the bridges are safe in general. Of course, if the risks are acceptable, this is also a strong indication that the safety measures for buildings are working.

5.4 Target Probability of Failure Values

Safety measures can only be applied if adequate target values are provided either in codes or at the state-of-the-art. Target values for the probability of failure and the safety index respectively are well known. Some target values are provided for further risk measures such as FAR as discussed before, however they are not as common in applications as the probability of failure.

Already in 1968 Freudenthal (1968) suggested a target probability of failure between 10^{-4} and 10^{-6} per year for steel structures and steel bridges and between 10^{-3} and 10^{-5} for reinforced concrete structures under design loads.

Since then, many references and codes such as Mathieu and Saillard (1974), Murzewski (1974), CEB (1976), CIRIA (1977), GruSiBau (1981), ISO 8930, DIN 1055-100 (1999), JCSS (2004) and the Eurocode (see also Spaethe 1992) give target values. In Tables 5.4, 5.5, 5.6, 5.7, 5.8, 5.9 and 5.10 examples are shown. Values close to 10^{-6} per year are shown in bold in those tables.

Zerna (1983) suggested probabilities of failure of 10^{-5} to 10^{-7} per year for the sudden loss of the equilibrium for a structure with large damage potential, of 10^{-4} per year for reaching the limit state without complete loss of load bearing capacity and between 10^{-2} and 10^{-3} per year for unsatisfactory behavior of the structure before reaching the load bearing capacity and with low damage consequences.

Menzies (1996) suggested an acceptable probability of failure for bridges in the range of 2×10^{-6} per year with a span less than 20 m.

Table 5.4 Target probably of failure per year according to the CEB (1976)

Average number of people endangered	Economical consequences		
	Low	Average	High
Low (<0.1)	10^{-3}	10^{-4}	10^{-5}
Average	10^{-4}	10^{-5}	**10^{-6}**
High (>10)	10^{-5}	**10^{-6}**	10^{-7}

Table 5.5 Target probability of failures in some Scandinavian countries (Spaethe 1992)

Safety class	Failure consequences	Probability of failure for the limit state of ultimate load per year
Low	Low personal injuries Insignificant economical consequences	1.0×10^{-4}
Normal	Some personal injuries Considerable economical consequences	1.0×10^{-5}
High	Considerable personal injuries Very high economical consequences	**1.0×10^{-6}**

Table 5.6 Target probability of failures in the former East Germany (Franz et al. 1991)

Reliability class	Consequences	Probability of failure
I	Very high danger to the public Very high economical consequences Disaster	1.0×10^{-7}
II	High danger to the public High economical consequences High cultural losses	**1.0×10^{-6}**
III	Danger to some persons Economical consequences	1.0×10^{-5}
IV	Low danger to persons Low economical consequences	1.0×10^{-4}
V	Very low danger to persons Very low economical consequences	7.0×10^{-4}

However most of the publications show nearly the same values: for new structures an annual maximum probability of failure in the range of 10^{-6} for the ultimate limit state and 10^{-3} per year for the serviceability limit state. These values are related to one limit state function such as bending, shear, buckling, normal forces etc., not to the overall structural probability of failure.

The Eurocode permits an adaptation of the safety index to consequences in terms of failure consequence classes (CC) as shown in Table 5.11. Such consequence

Table 5.7 Target probability of failures according to the GruSiBau (1981)

Safety class	Possible consequences of failure		Type of limit state	
	Limit state of ultimate load bearing	Limit state of serviceability	Ultimate load	Serviceability
1	No danger to humans an no economical consequences	Low economical consequences and low usage limitation	1.34×10^{-5}	6.21×10^{-3}
2	Some danger to humans and considerable economical consequences	Considerable economical consequences and strong limitation of further usage	$\mathbf{1.30 \times 10^{-6}}$	1.35×10^{-3}
3	High importance of the structure to the public	High economical consequences and high restriction to future usage	1.00×10^{-7}	2.33×10^{-4}

Table 5.8 Target probability of failure according to the DIN 1055-100 (1999) and the Eurocode 1 (1994)

Limit state	Probability of failure	
	Lifetime	Per year
Ultimate load	7.24×10^{-5}	$\mathbf{1.30 \times 10^{-6}}$
Serviceability	6.68×10^{-2}	1.35×10^{-5}

Table 5.9 Target safety indexes and target probabilities of failure (in brackets) according to ISO/CD 13822 (1999)

Limit state	Safety index (Probability of failure)
Serviceability	
Reversible	0.0
Irreversible	1.5
Fatigue	
Testable	2.3
Not testable	3.1
Ultimate load	
Very low consequences	2.3 (10^{-2})
Low consequences	3.1 (9.6×10^{-4})
Common consequences	3.8 ($\mathbf{7.2 \times 10^{-5}}$)
High consequences	4.3 (8.5×10^{-6})

Table 5.10 Target safety indexes and target probabilities of failure (in brackets) according to the JCSS Modelcode (2004)

Costs for safety measures	Low failure consequences	Average failure consequences	High failure consequences
Low	3.1 (9.6×10^{-4})	3.3 (4.8×10^{-4})	3.7 (1.1×10^{-4})
Average	3.7 (1.1×10^{-4})	4.2 (1.3×10^{-5})	4.4 (5.4×10^{-6})
High	4.2 (1.3×10^{-5})	4.4 (5.4×10^{-6})	4.7 (**1.3×10^{-6}**)

Table 5.11 Graduation of failure consequence classes according to the Eurocode 1 (1994)

Failure consequence classes	Consequences	Examples
CC 3	High consequences to humans, the economy. social systems and the environment	Stands, public buildings, for example concert halls
CC 2	Average consequences to humans, the economy, social systems and the environment	Dwelling and office buildings, public buildings such as offices
CC 1	Low consequences to humans. the economy, social systems and the environment	Agricultural structures or structures without regular persons residence, for example barns, conservatories

Table 5.12 Graduation of failure consequence classes according to the Eurocode 1 (Eurocode 1994)

Reliability class	Safety index per year (Probability of failure)	Safety index for 50 years (Probability of failure)
RC 3	5.2 (10^{-7})	4.3 (8.5×10^{-6})
RC 2	4.7 (**1.3×10^{-6}**)	3.8 (7.2×10^{-5})
RC 1	4.2 (1.3×10^{-5})	3.3 (4.8×10^{-4})

Table 5.13 Adaptation factor for the partial safety index subject to the Reliability class (Eurocode 1 1994)

Adaptation for the partial safety factors	Reliability class		
	RC1	RC2	RC3
K_{FI}	0.9	1.0	1.1

classes can then be related to some reliability classes (RC) listed in Tables 5.12 and 5.13. The reliability classes again show target values for the probability of failure.

Besides tables, also equations are known for target probabilities of failure. CIRIA (1977) document gives the following equation for the estimation of the target probability of failure

Table 5.14 Adaptation of the safety index according to the CAN/CSA-S6-88 Canadian Limit States Design Standard taken from Casas et al. (2001) and COST-345 (2004)

$\beta = 3.5 - (\Delta_E + \Delta_S + \Delta_I + \Delta_{PC}) \geq 2.0$	
Correction factor for element failure	Δ_E
Abrupt failure without warning	0.0
Abrupt loss of bearing capacity without warning with remaining capacity	0.25
Gratefully failure with warning	0.50
Correction factor for system failure	Δ_S
Failure of one single element causes system failure	0.00
Failure of one single element does not cause system failure	0.25
Failure of one single element causes local failure only	0.50
Correction factor for monitoring	Δ_I
Element is not controllable	−0.25
Element is controlled regularly	0.00
Critical elements are controlled more frequently	0.25
Correction factor for live load	Δ_{PC}
All types of traffic without special permission	0.00
All types of traffic with special permission	0.60

$$P_{ft} = \frac{10^{-4}}{n_r} K_s n_d$$

with P_{ft} as the probability of failure due to any cause during the design life n_d in years. n_r is the number of people at risk in the event of failure and K_s is given as 0.5 for bridges, for domestic and other buildings 0.05, for dams 0.005 and for towers and offshore structures 5. The estimated probabilities of failure are in the range of 10^{-6} to 10^{-5} per year.

The Eurocode furthermore can consider different types of production control of the building material in terms of changes of partial safety factors of the material. Still, this adaptation like all to other recommendations mentioned so far is proposed for new structures only.

However, some recommendations focus on existing structures. For example in Tables 5.14 and 5.15 some adaptation factors for the target safety index are given. Furthermore Strauss and Bergmeister (2005) have also introduced some factors. This adapted safety index can then be used to provide alternative safety measures in the semi-probabilistic safety concept (Fischer 2010; Fischer & Schnell 2008; Weber et al. 2018).

Allen (1992) suggested an increase of the safety index if a structure may fail suddenly without warning.

For existing structures usually less stringent requirements are common, for example a decrease of the safety index by 0.5 (Diamantidis et al. 2007, see also Sýkora

Table 5.15 Adaptation of the safety index according to Schueremans and Van Gemert (2001)

$\beta = \beta_T - (\Delta_S + \Delta_R + \Delta_P + \Delta_I) \geq 2.0$	
Adjustment for system behaviour:	Δ_S
Failure leads to collapse. likely to impact occupants	0.00
Failure is unlikely to lead to collapse. or unlikely to impact occupants	0.25
Failure is local only. very unlikely to impact on occupants	0.50
Adjustment for risk category:	Δ_R
High number of occupants (*n*) exposed to failure (*n* = 100–1.000)	0.00
Normal occupancy exposed to failure (*n* = 10–99)	0.25
Low occupancy exposed to failure (*n* = 0–9)	0.50
Adjustment for Past Performance:	Δ_P
No record of satisfactory past performance	0.00
Satisfactory past performance or dead load measured	0.25
Adjustment for Inspection:	Δ_I
Component not inspect able	−0.25
Component regularly inspected	0.00
Critical component inspected by expert	0.25

& Holicky 2013; Weber et al. 2018). Furthermore the probability of failure or the safety index can be adapted to more specific conditions.

Fischer (2010), Kotes & Vican (2012) and SIA 269 (2007) give a time-dependent target values for the safety index and the probability of failure (Fig. 5.4).

Rackwitz (1999) gives target safety indexes and probabilities of failure for structures at the end of their lifetime. The values are dependent on the cost of repair in the range of 2.2×10^{-2} and 1.6×10^{-1} per year. Vogel et al. (2009) refer further to fib recommendations giving slightly lower probabilities of failure, but significantly larger than 10^{-4} per year. These numbers have also been included in the Fig. 5.4.

Wang (2010a, b) states, that in the development of the AASHTO target values a safety index of 3.5 (2.33×10^{-4}) was selected for new bridges determined by calibration to a spectrum of traditional bridge design situations involving steel, reinforced concrete and pre-stressed concrete constructions. The value of 2.5 (6.2×10^{-3}) was chosen by judgement for existing bridges (see also Moses 2001). This indicates a difference of more than one order of magnitude between the target probability of failure for new and existing bridges.

Ghasemi (2015) suggests target safety indexes and probabilities of failure respectively depending on the span and the corrosion conditions for bridges.

Duckett (2005) suggests a risk of failure of 10^{-6} per year but an upper bound for bridges would be 10^{-4} per year. This 10^{-4} per year value is related to accidental loads such as ship collisions. Duckett (2005) refers for example to the AASHTO Guide specification and commentary for vessel collision design highway bridges which gives for critical bridges an acceptable annual frequency of collapse be equal

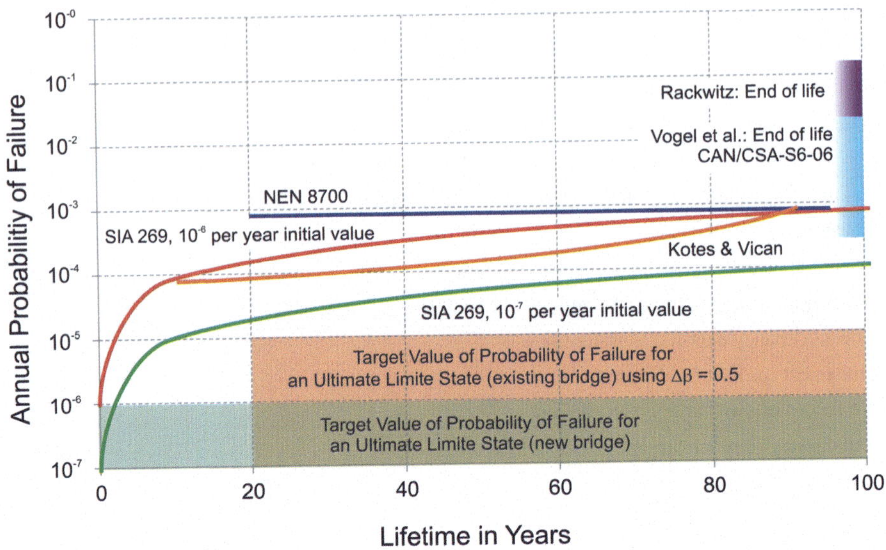

Fig. 5.4 Time dependent target probability of failure for existing bridges based on different references (Vogel et al. 2009; Kotes & Vican 2012; SIA 269 2007)

or less than 0.01 in 100 years giving an annual frequency of 10^{-4} and which gives for regular bridges an acceptable annual frequency of collapse be equal or less than 0.1 in 100 years giving an annual frequency of 10^{-3}. Duckett (2005) also refers to IABSE Structural Engineering Documents from 1993 which define the representative value of the accidental action to be chosen in such a way that there is an assessed probability less than 10^{-4} per year for one structure that a specific or a higher energy will occur. However this is not a probability of failure but a probability of a load.

MacDonald et al. (2016) gives acceptable probabilities of failure for bridges during military actions. He provides the following equation:

$$P_{fyr} = 1 - \left(1 - \frac{n_d}{d_p}\right)^{\frac{1\,\text{year}}{T}}$$

with P_{fyr} is the probability of fatality per year, n_d denotes the number of military fatalities in the conflict, n_p is the total number of military personal involved in the conflict. T is the duration of the conflict in years.

Figure 5.5 visualises the acceptable risk for members of the army compared to the probability of being killed by bridge crossings. We can see that the risk of being killed due to bridge crossing is about two orders of magnitude lower then the acceptable general risk to members of the army during different conditions. This is the same ratio used in the Netherlands with an acceptable risk of 10^{-6} per year for the public and a general risk of 10^{-4} per year.

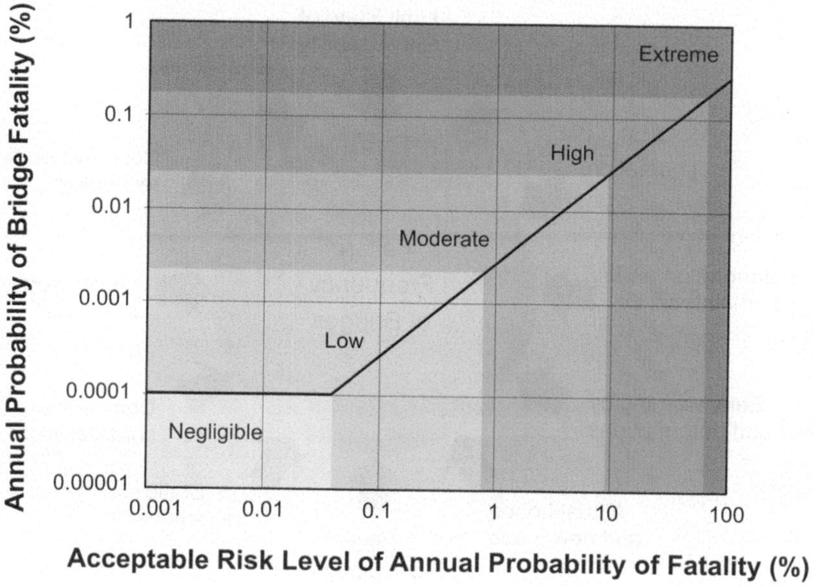

Fig. 5.5 Acceptable annual risk for military bridge crossings (MacDonald et al. 2016)

The fast majority of the codes, documents, guidelines and scientific references recommend a probability of failure for a ultimate limit state of 10^{-6} per year. This value reminds one very much to the 10^{-6} de-minimis-risk for the public. The de-minimis-risk is a regulatory cut-off for risks which are not part of codes and guidelines since they are too low: they are acceptable (Proske 2009).

The 10^{-6} value is widely used not only in structural engineering, but elsewhere, for example in risk management for chemicals, where 10^{-6} as additional risk of cancer is known.

5.5 Correction of Probability of Failure

5.5.1 Introduction

The target probability of failure considers a limited number of effects such as the random variables, the mechanical behaviour and model respectively and the equation of proof.

An example of the limitation is the fact that the target value for the probability of failure of structures is related to one single limit state. Therefore we have to convert the value to the probability of failure for the entire structure since the safety of the structure depends on many limit states related to various structural elements.

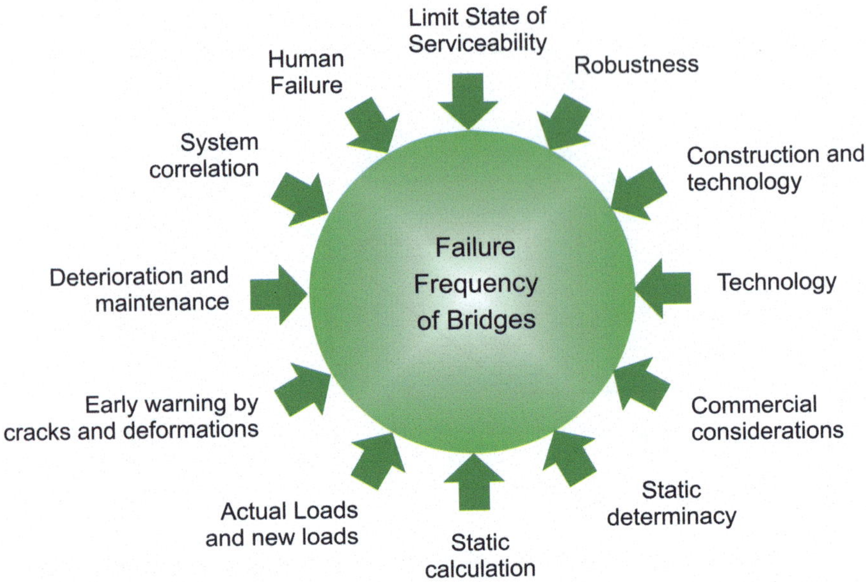

Fig. 5.6 Potential factors influencing the observed failure frequency of bridges

Only the structural probability of failure can be compared with an observed collapse frequency! The conversion has to be carried out by the consideration of all structural elements with their limit states. Furthermore, as stated in different publications and codes, the target value does not consider correlations between the various limit state functions and the different structural elements which are part of the structure. It does also not consider human failure.

Figure 5.6 lists a variety of factors, which are essentially not considered in the calculation of the probability of failure. All these factors can be considered by a correction. The correction can be formulated as:

$$P_{f\,\text{Structure}} = \frac{P_{f\,\text{Limit State}} \times \prod F_{\text{negative}}}{\prod F_{\text{positive}}}$$

with $P_{f\,\text{Structure}}$ as probability of failure of the structure, $P_{f\,\text{Limit State}}$ as probability of failure of the single limit state, F_{negative} as factors not considered in the computation of the $P_{f\,\text{limit State}}$ yielding to an increase of the overall probability of failure and F_{positive} considering factors decreasing the $P_{f\,\text{limit State}}$. Of course, some software packages are able to compute the probability of failure for the structure directly.

However, for the sake of simplification and as example, we can consider the correlation and human failure as:

$$P_{f\,\text{Structure}} = \frac{P_{f\_\text{Limite State}} \times HF}{C}$$

with HF as overall Human Failure factor for the entire structure and C as Correlation factor for the entire structure.

In the following section individual correction parameters are discussed.

5.5.2 Correlation Consideration

The definition of target values of the probability of failure for limit states is correct, otherwise the target values have to consider specific information about the individual structure. For example a structure with many columns would require another target value as a structure with a low number of columns. Also the technology of the construction of the columns or the number of building enterprises involved in the construction may influence the outcome. Some of such information must then be used by the designer, but usually this information is unknown to the designer.

Usually the construction technique is comparable within one structure and the number of building enterprises in one building is limited. Therefore we can assume a high correlation between the many limit states and we can discard the information. This simplification means, if one column fails, probably all columns would fail under the same loading.

Tanner and Hingorani (2015) assume that the independence of the limit states is reasonable in most cases especially for buildings with statically determined structures. Spaethe (1992) indicates that the limit states in structures and therefore also in bridges are never independent even if the random variables itself are not correlated since several random variables are used in different limit state functions. For example, loads are considered for different limit states functions in one structural element.

However in cases, in which the structural system shows many different and non-correlated failure modes, the target of the probability of failure should be decreased (Rackwitz 1998).

Indeed, the Eurocode permits the adaptation of the target probability of failure related to some reliability and consequence classes as mentioned before.

5.5.2.1 System

Besides the simplified approach by the Eurocode, the overall probability of failure can be directly computed for systems of structural elements. This can be done either by defining various limit state function in the computation of the probability of failure in one random variable space or by combining different probabilities of failure using logical combinations. Such combinations have been provided for a variety of systems such as parallel or series systems.

The effect of correlation depends on the number of individual elements, the logical combination of the individual elements and the amount of correlation. If in a parallel system the individual elements are independent, the system failure probability is the

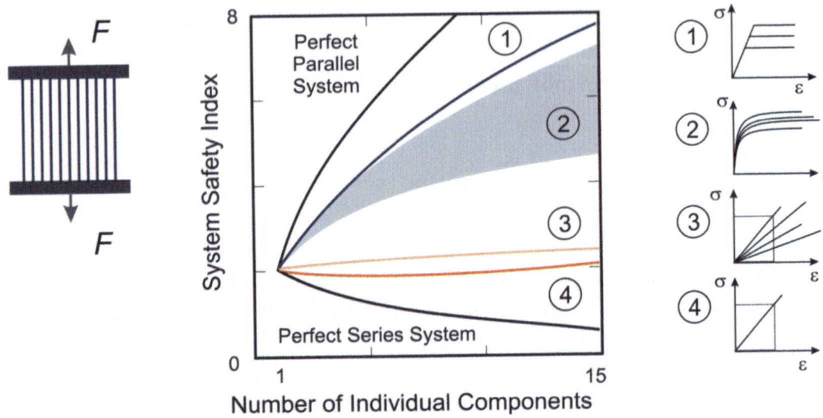

Fig. 5.7 Visualisation of the dependency of the system safety index and the number of individual components with different stress-strain-relationships (Gollwitzer and Rackwitz 1990)

product of the individual failure probabilities. One can well imagine that the overall failure probability decreases and the overall safety index increases with every further non-correlated element (see Fig. 5.7). In contrast, if the elements form a series system, the overall failure probability increases and the overall safety index decreases with every element since it is the sum of all individual failure probabilities (see Fig. 5.7). In the first case, the correlation will increase the overall probability of failure, in the second case the overall failure probability will decrease with increasing correlation. A structure is usually a series system, if one component fails, the entire structure fails. However, this fact is not entirely true for static indeterminate structures. Here, if a local failure occurs, the structure will not collapse. Figure 5.7 shows the overall safety index for a Daniels-system (left) considering parallel systems, series systems and elements with different stress-strain-relationships.

As mentioned, series systems are extremly important for structural systems. Most of the structural systems are series systems because if one column or one beam fails, at least parts of the structure will collapse. Therefore, the neglect of correlation leads to conservative results.

Additional to the approach for standard combinations of elements (parallel, series system) some authors (Cornell 1967; Ditlevsen 1979; Thoft-Christensen and Baker 1982; Greig 1992; Rackwitz 2001; Madsen et al. 2006) have developed simplified rules to provide bounds and to combine individual probabilities of failure for situations common in structural engineering and considering realistic correlations (see Spaethe 1992; Voigt 2014).

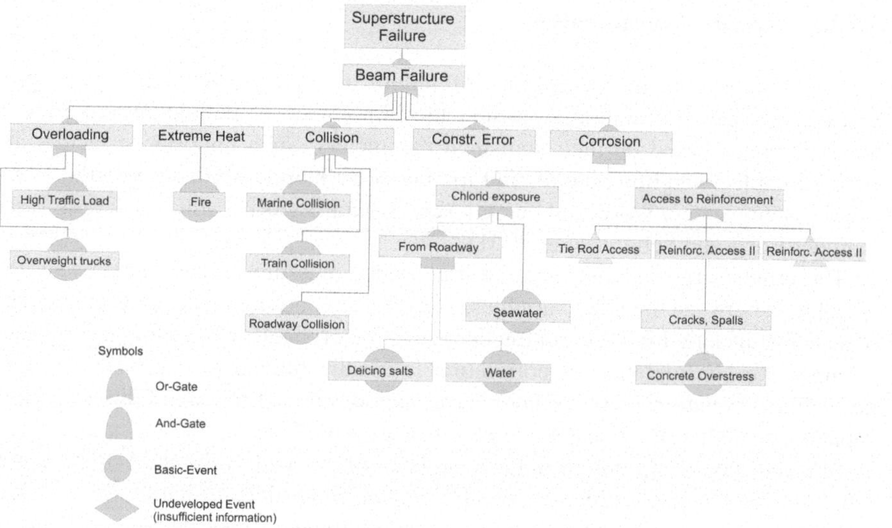

Fig. 5.8 Example of a Fault Tree (example taken from Davis-Mcdaniel 2011)

5.5.2.2 Fault Trees and Event Trees

Fault Trees or Fault Tree Analysis are methods to combine the failure probability of single elements or parts of a system into an overall system probability of failure. The combination can consider the logical construction of the overall system and therefore providing a realistic result for the system. The individual parts can for example include the failure of a pump starting or running or different failures of a switch. Therefore Fault Trees can become very large and often, they are sub-divided. The results of Fault Trees can be included into Event Trees. Event Trees model the sequence of events or accidents. For example, they can describe, what happens, if a pump or a switch fails and which further failures are required that a safety goal will finally not be reached.

The application of such techniques is common in Nuclear Engineering, Chemical Industries, Electrical Engineering and other fields. The application is not common in structural engineering, however first applications of the Boolean approach for structures can be found in Helbig (1987).

Application examples for bridges are known such as Fischer et al. (2004), LeBeau and Wadia-Fascett (2007), Davis-Mcdaniel (2011), Straub and Der Kiureghian (2011), Müller et al. (2011) and Davis-Mcdaniel et al. (2013). The example in Fig. 5.8 is taken from Davis-Mcdaniel (2011) and shows a sub-tree. The overall Fault Tree covers more elements.

5.5.2.3 Specific Consideration

The computation of the probability of failure as well as the safety index provides additional information regarding the importance of individual variables in the limit states based on their statistical parameters. Such importance factors are either so-called weighting factors in First Order Reliability Method (Spaethe 1992) or Risk Achievement Worth and Fussell-Vesely-parameters in Fault- and Event-Trees (van der Borst and Schoonakker 2001).

The probability of failure and the importance measures depend heavily on the correlation between the input variables. The effect of correlation depends on the type of systems; usually high correlation decreases the probability of failure in bridges.

Since the consideration of correlation is usually limited in Fault-Trees (either no or full correlation) new methods have been developed to extend the modelling capabilities such as the balancing method (Kim et al 2005).

In recent decades, random fields have been introduced in the field of structural reliability, which also represent a specific model for spatial correlation.

Finally one has to keep in mind that the empirical correlation values converge only very slow. In other terms, large sample sizes are required to achieve robust and converging empirical correlation values. This fact is shown in Fig. 5.9. One can easily see that a large number of samples are required to reach an acceptable small confidence interval for the correlation factor.

5.5.2.4 Conclusion

The consideration of correlation is very challenging since the estimation of correlation of the same type of structural elements and different types of structural elements may depend on many different conditions: the same workers on site, the same concrete supplier and so on. All these information is unknown.

Furthermore the mathematical consideration of correlation in Fault Trees can be tricky since in many cases methods such as the Multi-Greek-Letter-method were developed for mechanical systems with either full or no correlation. These methods may not be applicable for structural systems.

On the other hand, several techniques have been developed to combine single limit state probabilities to structural probabilities.

5.5.3 Human Error Consideration

Human error is a common part of all human actions. In general human error is a deviation from an intention caused by human action.

Table 5.16 shows the fraction of human error occurrence during design, construction and use of structures. Most human errors occur during the design and the

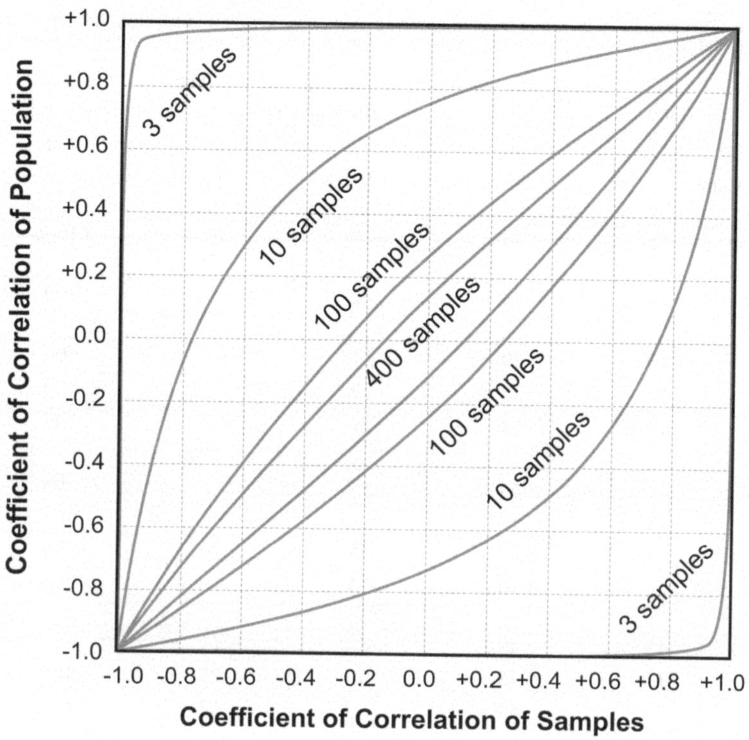

Fig. 5.9 Ninety-five percent confidence interval for coefficient of correlations (Steel and Torrie 1991)

construction of structures and bridges. Table 5.17 shows in detail human error rates related to certain micro-tasks in the design and construction process.

Table 5.18 and Figs. 5.10, 5.11 and 5.12 give causes of human errors related to structural failures and damages. Table 5.18 is an extension of the database behind Fig. 5.10. The table confirms the conclusions by Fig. 5.10, that carelessness and insufficient knowledge are major contributors for structural collapses. Real mistakes are contributing only minor. Figure 5.11 which is related to damages found on structures confirms Table 5.16, stating that faulty design and inappropriate constructions are major contributors. Insufficient preliminary investigation can be interpreted as carelessness.

These facts are well known and several quality assurance procedures and independent control and checking of all steps are part of the design and construction process. By such procedures the error rate during design and construction can be decreased at least by one order of magnitude, often by nearly two orders of magnitude (Stewart 1993).

However, as structural collapses show, such errors still exist. Therefore they have to be considered in the estimation of the structural probability of failure.

Table 5.16 Percentage of human errors made in certain phases of structures (taken from El-Shahhat et al. 1995; Fröderberg 2014)

Reference	Planning and design	Construction	Use and maintenance	Others	Total
CEB 157	50	40	8	–	98
Matousek and Schneider	37	35	5	23	
Brand and Glatz	40	40	–	20	100
Yamamoto and Ang	36	43	21		100
Grunau	40	29	31		100
Reygaerts	49	22	29		100
Melchers et al.	55	24	21		100
Fraczek	55	53	–		108
Allen	55	49	–		103
Hadipriono	19	27	33	20	99
Rackwitz and Hillemeier	46	30	23	1	100
Matousek	45	49	8	2	
Hauser	37	35	5	23	100
Gonzales	29	59	–	12	100

Liu (2000) gives variations of the probability of failure by up to 25% based on changes of cross sections by human errors. On large, statically indetermined structures, this value will be lower. Therefore a variation of 5–10% seems to be a reasonable selection.

In other cases, simulations showed several orders of magnitude difference between the probability of failure for error-free and error-including structures (Fröderberg 2014). One of these cases was related to misunderstanding regarding loads (see Fig. 5.11).

In general, one can always argue that human error is responsible for all collapses of bridges. This has been stated several times as shown in Chap. 2. Another example of this theory is taken from Tarkov (1986): "Honest human error in the face of the unforeseen—or the unforeseeable—is ultimately what brings bridges down." Tweed (1969) has introduced a classification of ignorance causing bridge failure.

If we assume that bridge collapses are always related to human error, we have to install adequate mitigation measures to provide sufficient safety for humans and societies. Such an approach is the concept of safety culture. It is based on the assumption that disasters do not happen by a single individual human failure, but by a cascade of failures. Reason (1990) has introduced such a model consisting of different layers or barriers (Fig. 5.13). Sometimes it is called the Swiss Cheese Model, where a disaster

Table 5.17 Micro-task error rates according to different references related to phases design and construction

Micro-task	Error rate	Reference
Design		
Code interpretation	0.015	Stewart and Melchers (1989)
Chart look-up	0.020	Beeby and Taylor (1973)
Table look-up	0.013	Melchers (1984)
One-step calculation (e.g. a × b)	0.013	Melchers (1984)
Two-step calculation (e.g. a × b – c)	0.026	Melchers (1984)
Three-step calculation	0.038	Stewart and Melchers (1989)
Four-step calculation	0.051	Stewart and Melchers (1989)
Five-step calculation	0.064	Stewart and Melchers (1989)
Eight-step calculation		Stewart and Melchers (1989)
Self checking of calculation[a]	0.90	Stewart and Melchers (1989)
Independent checking of calculation[a]	0.65	Stewart and Melchers (1989)
Construction		
Reduced number of reinforcement bars	0.022	Stewart (1993)
Increased number of reinforcement bars	0.011	Stewart (1993)
Decrease of section height	0.019	Stewart (1993)
Increase of section height	0.030	Stewart (1993)
Decrease of section width	0.008	Stewart (1993)
Increase of section width	0.008	Stewart (1993)
Insufficient concrete strength	0.22	Liu (2000)

[a]This value is related to overlook of an error in the calculation

Table 5.18 Causes of errors in the design and construction of building structures (Bea 1994)

Reference	Negligence, carelessness in %	Insufficient knowledge in %	Mistakes in %	Reliance on others in %	Other sources in %
Matousek (1982)	35	38	9	6	12
Melchers (1984)	24	52	8	2	13
Eldukair and Ayyub (1991)	82	67		29	33–72

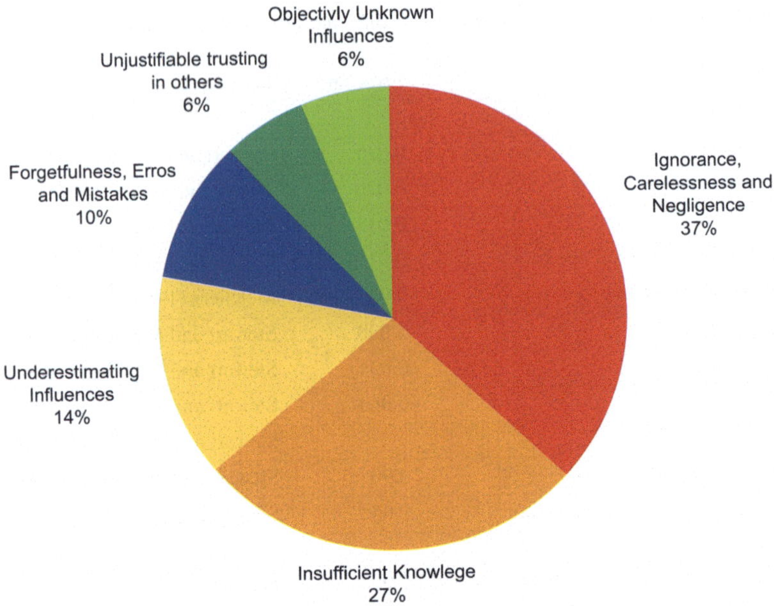

Fig. 5.10 Causes of structural failures according Matousek and Schneider (1976)

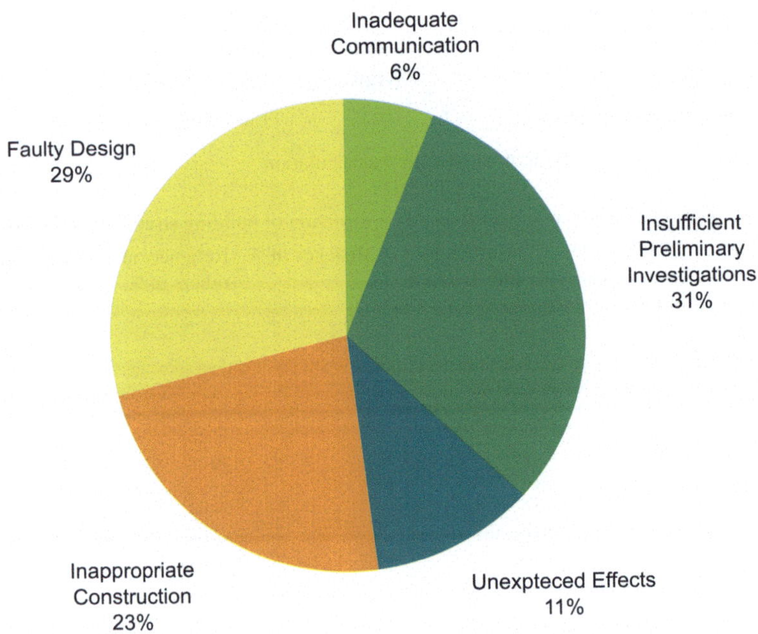

Fig. 5.11 Causes of structural damages related to steel sheet piling (Rizkallah et al. 1990)

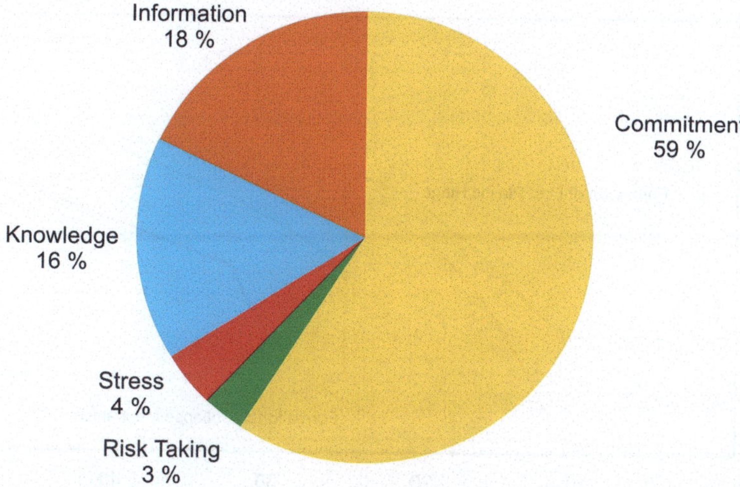

Fig. 5.12 Causes of faults of structural systems (Josephson and Hammarlund 1996)

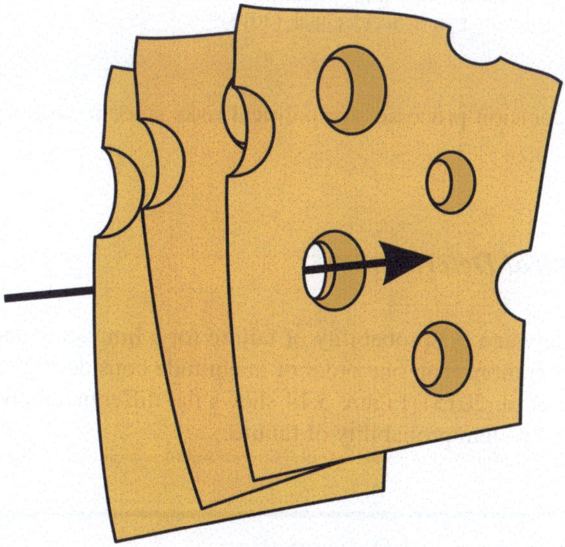

Fig. 5.13 Reason's (1990) Swiss Cheese Model (A disaster is a sequence, which penetrates all layers. Layers are safety measures.)

only occurs, if all Cheese levels are penetrated by a sequence. The concept of safety culture is strongly related to the experience gained from large disasters.

This approach is in compliance with the Integral Risk Management Approach for natural risks (Kienholz et al. 2004), Living Safety Analysis for technical risks or

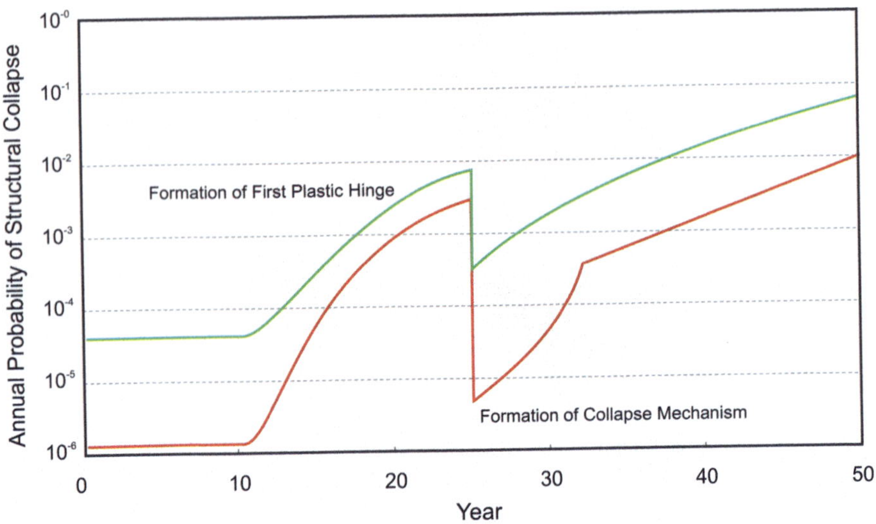

Fig. 5.14 Probability of failure for a bridge related to the formation of the first plastic hinge and the collapse mechanism based on Schneider et al. (2015)

Risk Informed Decision processes for political risks since it assumes a continuous improvement.

5.5.4 *Structural Determinacy*

The difference between the probability of failure for a limit state and for the entire structure can reach more than one order of magnitude considering statical determinacy (Schneider et al. 2015). Figure 5.14 shows the difference between the single limit state and the system probability of failure.

5.5.5 *Maintenance and Deterioration*

Maintenance and deterioration are two processes which can substantially affect the probability of failure of structures. Usually, the maintenance decreases the probability of failure in a short period, e.g. an event, whereas the deterioration increases the probability of failure over a long period. Figures 5.14 and 5.15 show examples illustrating both processes. Figure 5.14 does not only show the development of the probability of single hinge development, but also for the entire structure in terms of formation of a collapse mechanism.

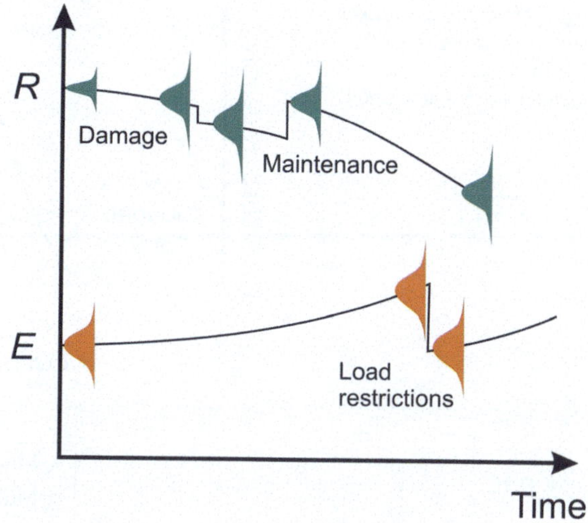

Fig. 5.15 Concept of probability of failure over time related to maintenance and deterioration

Figure 5.15 shows the general concept of maintenance and deterioration, which has been used widely in Life Cycle Cost Optimisation. Besides the overall concept, the figure also indicates that maintenance measures can unintentionally also yield to an increased probability of failure.

5.5.6 Actual Loads and New Loads

Codes use simplified models of the observed and assumed traffic and live loads. Background information regarding the road and railway traffic models can be found in Proske and van Gelder (2009) and Proske and Loos (2009).

For example, it is well known that weight restrictions are not fully met by drivers. Figure 5.16 shows the concept of overall vehicle weights indicating that weight limits (Q_{perm}) are exceeded substantially. Figure 5.17 shows vehicle weight measurement results from the Blue Wonder Bridge in Dresden, Germany which is limited to 15 tonnes. In contrast, the measurement indicates weights up to 21 tonnes.

On the other hand the adaptation of the weight limit to the bridge conditions is an effective and efficient way to decrease the probability of failure for bridges under live and dead load. This is illustrated in Fig. 5.18 which shows the frequency of weight restricted traffic (15 tonnes) on the Blue Wonder Bridge in comparison to the weights measured on a highway bridge. The latter is the basis for the road traffic model in the Eurocode.

Therefore weight restrictions, lane restrictions and speed limits are often used to correct the loading conditions to the bridge conditions.

In contrast, developments in other technical fields which can not be foreseen by the design engineer can negatively influence the probability of failure of bridges. For

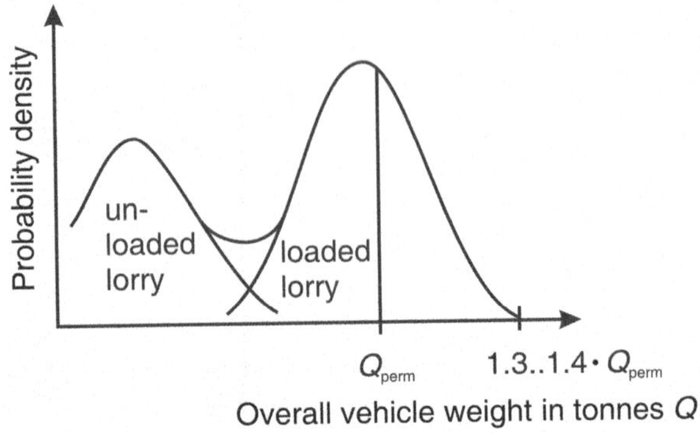

Fig. 5.16 General bimodal random distribution of the overall vehicle weight (Geißler 1995; Quan 2004)

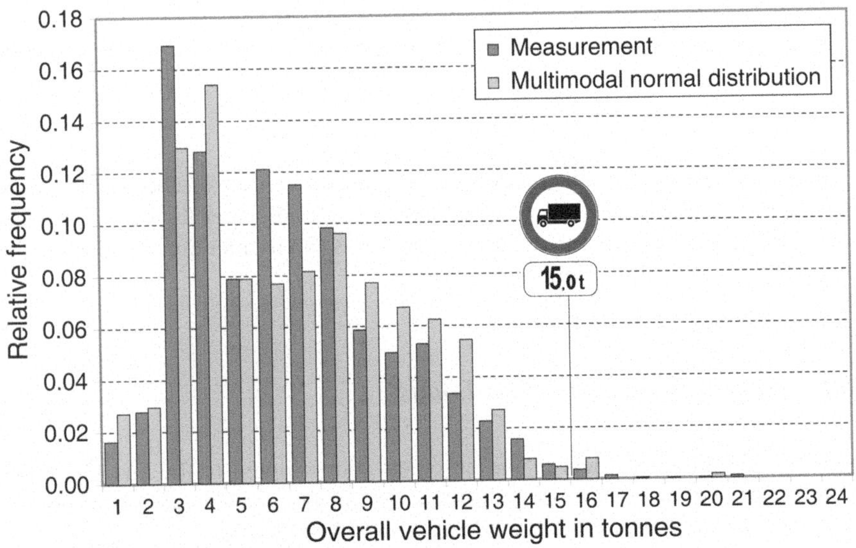

Fig. 5.17 Relative frequency of measured overall vehicle weight and adjusted multimodal normal distribution of heavy weight vehicles in October 2001 at the Blue Wonder Bridge in Dresden (Proske and van Gelder 2009)

example new traffic control systems may significantly change the traffic between different routes and may increase the load on specific bridges, also new types of cars and lorries (in Germany larger trucks called Monster trucks with an overall length of 25.25 m have been permitted recently) or new types of ships due to enlargement of channels may effect the bridge loads. The opening of the Rhine-Main-Danube channel in 1992 yielded to larger ships and to higher risks of bridge failure due to ship impact. It was later shown that 55 bridges on the river Main were not safe under ship impact (Mainpost 2009).

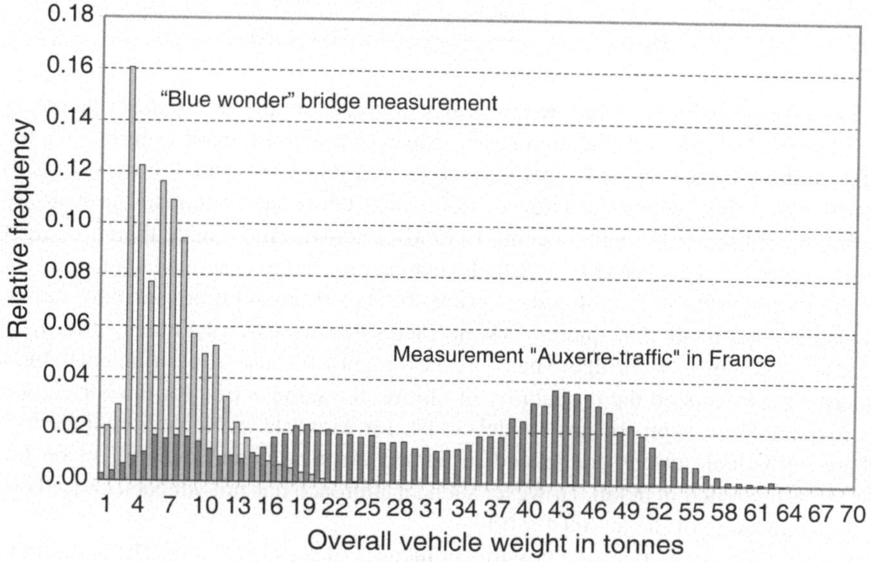

Fig. 5.18 Comparison of the overall vehicle weights measured at the Blue Wonder Bridge in Dresden and at the Auxerre-traffic in France

5.5.7 Structural Probabilities of Failure

In this section as an example the probability of failure is corrected by considering two additional effects only, correlation between the structural elements and human error probabilities. We are able to compute a structural probability of failure according to:

$$P_{f\ \text{Structure}} = \frac{P_{f\_\text{Limite State}} \times HF}{C}$$

with HF as overall Human Failure factor for the entire structure and C as Correlation factor for the entire structure.

In the example we assume:

$$P_{f\ \text{Structure}} = \frac{10^{-6} \times 1.05}{0.80} = 1.3 \times 10^{-6}$$

This example would correspond to an increase of the structural probability of failure by about 30% and a decrease of the safety index by about 2%. One of the reasons of the introduction of the safety index was the improved handling and the robustness. This becomes very clear in our example.

However if the human error rate is increased to about 30%, the structural probability becomes:

$$P_{f\,\text{Structure}} = \frac{10^{-6} \times 1.30}{0.80} = 1.63 \times 10^{-6}$$

Then the probability of failure has nearly doubled and the safety index increases by 3 to 4%. As seen in the section target values of probabilities of failure (5.4), the safety index of existing structures may be decreased by 0.5 which corresponds to nearly 10%. That means the largest discrepancy between the realistic probability of failure and the target value occurs right after construction, since then the target value for new bridges has to be fulfilled whereas after some time the target value for existing bridges has to be fulfilled covering effects as human failure and correlation.

On the other hand, the equation is simplified since one essential human error may yield to the collapse of a bridge. Therefore the human error rate is not sufficient to fully describe the effects on the probability of failure; the same is true for the correlation coefficient. More sophisticated models exist, for example in Nuclear engineering, where individual actions are related to human error rates and their effect on the control is modelled in detail. However such an approach has not yet been carried out to the knowledge of the author for bridges.

Figure 5.19 summarizes the possible influences of certain factors. The magnitude of the factors is estimated based on expert judgment and subjective experience. Some of the factors can indeed largely affect the probability of failure and thereby the observed collapse frequency. However, several of the factors are both, positive and negative. Considering the central limit theorem of the statistics, the sum and product (logarithm) of a large number of individual random numbers will yield to a normal distribution independent of the distribution type of the single factors. Therefore the probability of failure will be multiplied with a normal distribution with a mean value around zero. Hence the probability of failure should not significantly change over the bridge population. This will be discussed in the next chapter.

5.6 Conclusion

Based on the discussion in this chapter, we conclude that the probability of failure is the most adequate measure of safety because

- there are tools for the computation of the probability of failure and the computation is state-of-the-art,
- an extensive scientific discussion about target values for limit states has been carried out in the last decades, target values have been defined (consent of values) and are applied,
- a conversion of the target values from the limit state to the entire structure is possible and methods are available,

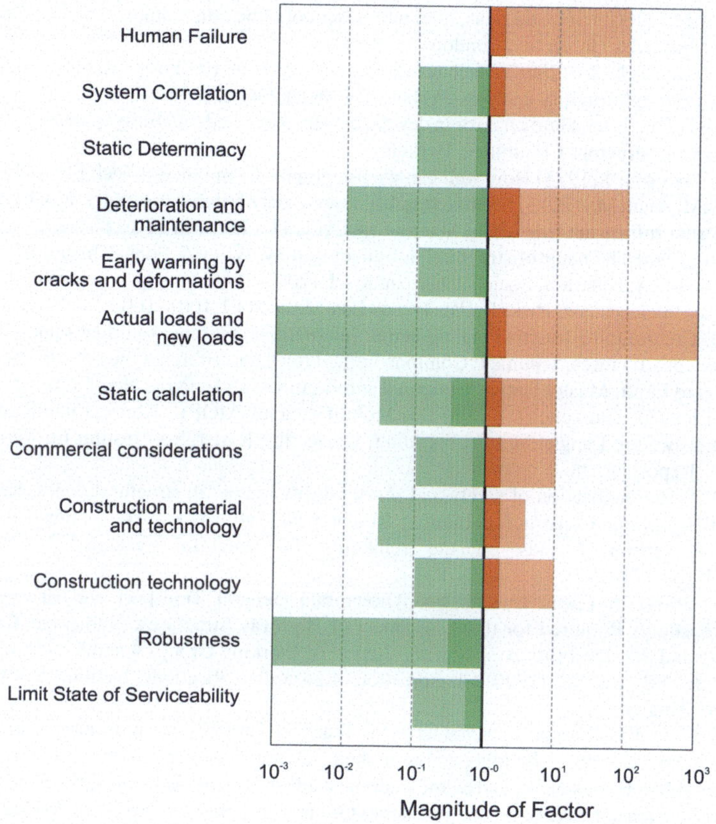

Fig. 5.19 Tornado plot of the possible influences on the probability of failure and the collapse frequency

- the probability of failure of the entire structure can indeed be compared to the frequency of collapse if further effects such as human failure, correlation and others are considered and
- further scientific discussion is required regarding the target values of other risk measures, although individual applications can be found.

References

Allen DE (1992) Canadian highway bridge evaluation: reliability index. Can J Civ Eng 19:593–602

Aven T, Vinnem JE, Vollen F (2005) Perspectives on risk acceptance criteria and management for installations—application to a development project. In: Kolowrocki (ed) Advances in safety and reliability, Taylor & Francis Group, U.K, pp 107–114

Ball DJ, Floyd PJ (2001) Societal risks, final report, school of health, biological and environmental sciences. Middlesex University, London

Banerjee S, Shinozuka M (2008) Mechanistic quantification of RC bridge damage states under earthquake through fragility analysis. Probab Eng Mech 23(1):12–22

Bea RG (1994) The Role of human error in design, construction, and reliability of marine structures, SSC-378. Ship Structure Committee, Berkeley

Beeby AW, Taylor HPJ (1973) How well can we use graphs. Commun Sci Tech Inf 17:7–11

Billah AHMM, Alam MS (2013) Seismic fragility assessment of highway bridges: a state-of-the-art review. Struct Infrastruct Eng Maint Manag Life Cycle Des Perform 11(6):804–832

Blockley DI (1980) The natur of structural design and safety. Wiley & Sons, Chichester

Casas JR, Prato CA, Huerta S, Soaje PJ, Gerbaudo CF (2001) Probabilistic assessment of roadway and railway viaducts. Saf, Risk, Reliab Trends Eng Malta 2001:1001–1008

CEB (1976) Comité Euro-International du beton: International system of unified standard—codes of practice for structures, Volume I: Common unified rules for different types of construction and material (3rd Draft, Master Copy), Bulletin d'information 116 E, Paris, Nov 1976

Chlond B, Lipps O, Zumkeller D (1998) Das Mobilitätspanel (MOP)—Konzept und Realisierung einer bundesweiten Längsschnittbeobachtung, Universität Karlsruhe—Institut für Verkehrswesen, IfV—Report Nr. 98–2

CIRIA (1977) Rationalisation of safety and serviceability factors in structural codes, Report No. 63, Seminar Edition, Construction Industry Research and Information Association, London

Cornell CA (1967) Bounds on the reliability of structural systems. J Struct Div 93(ST1/1967):171–200

COST-345 (2004) European Commission Directorate General Transport and Energy: COST 345—Procedures Required for the Assessment of Highway Structures: Numerical Techniques for Safety and Serviceability Assessment—Report of Working Groups 4 and 5

Cox AW, Lees FP, Ang ML (1990) Classification of hazardous locations. Institution of Chemical Engineers, Rugby

Daniel JE, Wenzel F, Werner A, Schaefer AM, Tsang HH (2017) An introduction of tools for evaluating the influence of building codes on the level of earthquake fatalities globally, In: Hao, Zhang (eds) Mechanics of structures and materials: advances and challenges, Proceedings of the 24th Australasian conference on the Mechanics Of Structures and Materials (ACMSM24), Perth, Australia, 6–9 Dec 2016, Taylor & Francis Group, London, pp 1633–1639

Das PC (1997) Safety of bridges. Thomas Telford, London

Davis-Mcdaniel C (2011) Fault-Tree model for bridge collapse risk analysis. Thesis, Clemson University, MSc

Davis-Mcdaniel C, Chowdhury M, Pang W, Dey K (2013) Fault-Tree model for risk assessment of bridge failure: case study for segmental box girder bridges. J Infrastruct Syst 19(3):326–334

Der Kiureghian A, Haukaas T, Fujimura K (2006) Structural reliability software at the University of California, Berkeley. Struct Saf 28:44–67

Diamantidis D, Holicky M, Jung K (2007) Assessment of existing structures—on the applicability of the JCSS recommendations, In: Aspects of Structural Reliability—In Honor of R Rackwitz. In: Faber Vrouwenvelder (ed) Zilch. Herbert Utz Verlag, München

DIN 1055–100 (1999) Einwirkungen auf Tragwerke, Teil 100: Grundlagen der Tragwerksplanung. Sicherheitskonzept und Bemessungsregeln, Juli, p 1999

Ditlevsen O (1979) Narrow reliability bounds for structural systems. J Struct Mech 7(4/1979):453–472

Duckett W (2005) Risk analysis and the acceptable probability of failure. Struct Eng 83(15):25–26

Eldukair ZA & Ayyub BM (1991) Analysis of Recent U.S. Structural and Construction Failures. J Perform Constr Facil 5(1): 57–73, February

El-Shahhat AM, Rosowsky DV, Chen WF (1995) Accounting for human error during design and construction. J Archit Eng 1(2):84–92

EPRI (1994) Methodology for developing seismic fragilities, prepared by J. R, Benjamin and Associates, Inc and RPK Structural Mechanics Consulting, TR-103959, Project 2722-23, June 1994

Epstein S, Rauzyb A, Reinhart FM (2008) The open PSA initiative for next generation probabilistic safety assessment. In: Schmidt H, Proske D (eds) Proceedings of the 6th International Probabilistic Workshop, Graubner, Darmstadt

Eurocode 1 (1994) (ENV 1991–1): Basis of design and action on structures, Part 1: Basis of design, CEN/CS, Aug 1994

Fischer A (2010) Bestimmung modifizierter Teilsicherheitsbeiwerte zur semiprobabilistischen Bemessung von Stahlbetonkonstruktionen im Bestand, Dissertation, Kaiserslautern

Fischer A, Schnell J (2008) Determination of partial safety factors for existing structures. In: Graubner CA, Schmidt H, Proske D (eds) Proceedings of the 6th International Probabilistic Workshop, Darmstadt, pp 133–147

Fischer J, Straub D, Schneider R, Thöns S, Rücker W (2004) Intelligente Brücke—Zuverlässigkeitsbasierte Bewertung von Brückenbauwerken unter Berücksichtigung von Inspektions- und Überwachungsergebnissen, Bundesanstalt für Straßenwesen, Heft B99, Bergisch-Gladbach

Franz G, Hampe E, Schäfer K (1991) Konstruktionslehre des Stahlbetons, Band II: Tragwerke, Zweite Auflage, Springer Verlag, Berlin

Freudenthal AM (1947) Safety of structures. Trans ASCE, V 112(1947):127–180

Freudenthal AM (1968) Critical appraisal of safety criteria and their basic concepts. IABSE congress report 8(1968):13–24

Fröderberg M (2014) The human factor in structural engineering—a source of uncertainty and reduced structural safety. Division of Structural Engineering, Faculty of Engineering, Lund University, Sweden

Geißler K (1995) Beitrag zur probabilistischen Berechnung der Restnutzungsdauer stählerner Brücken, Dissertation, Heft 2 der Schriftreihe des Institutes für Tragwerke und Baustoffe an der Technischen Universität Dresden

Ghasemi SH (2015) Target reliability analysis for structures. Dissertation, Auburn University, May 2015

Gollwitzer S, Rackwitz R (1990) On the reliability of Daniels systems. Struct Saf 7:229–243

Gollwitzer S, Kirchgäßner B, Fischer R, Rackwitz R (2006) PERMAS-RA/STRUEL system of programs for probabilistic reliability analysis. Struct Saf 28:108–129

Greig GL (1992) An Assessment of high-order bounds for structural reliability. Struct Saf 11:213–225

GruSiBau (1981) Normenausschuß Bauwesen im DIN: Grundlagen zur Festlegung von Sicherheitsanforderungen für bauliche Anlagen, Beuth Verlag

Haldi PA, Vulliet L (1998) Fiabilité et sécurité des systèmes civils. Lecture notes, Swiss Federal Institute of Technology (EPFL), Lausanne

Helbig S (1987) Systemzuverlässigkeit von Konstruktionen im Maschinenbau und Bauwesen auf der Theorie der monotonen Systeme. Akademie der Wissenschaften, Institut für Mechanik, Karl-Marx-Stadt

Hingorani R (2017) Acceptable Life Safety Risks associated with the effects of Gas Explosions on Reinforced Concrete Structures, Dissertation, Universidad Politecnica De Madrid

ICOLD (2005) Risk assessment in dam safety management, ICOLD Bulletin 130

ISO 2394 (1998) General principles on reliability of structures, 2nd ed. Springer, New York

ISO, CD 13822 (1999) Bases for design of structures—Assessment of existing structures. International Organisation for Standardization, Geneva

JCSS (2004) Joint committee of structural safety, Probabilistic Modelcode, www.jcss.ethz.ch

Jonkman SN (2007) Loss of life estimation in flood risk assessment, theory and applications. Dissertation, Delft University of Technology, Delft, the Netherlands, 360 pp

Jonkman SN, van Gelder PHAJM, Vrijling JK (2003) An overview of quantitative risk measures for loss of life and economic damage. J Hazard Mater A99:1–30

Josephson PE, Hammarlund Y (1996) Kvalitetsfelkostnader in the 90s—A study of seven building projects, Part I: Results, Gothenburg: Report 49th, Chalmers University, Inst. Construction Economy
Kennedy RP (1999) Overview of methods for seismic PRA and SMA Analysis including recent innovations. In: Proceedings of the OECD/NEA workshop on seismic risk, 10–12 Aug 1999, Tokyo, Japan
Kennedy RP, Cornell CA, Campbell RD, Kaplan S, Perla HF (1980) Probabilistic seismic safety of an existing nuclear power plant. Nuclear Eng, Design 59:315–338
Kienholz H, Krummenacher B, Kipfer A, Perret S (2004) Aspect of integral risk management in practice—considerations. Österreichische Wasser- und Abfallwirtschaft. vol 56, Heft 3–4, Mar–April 2004, pp 43–50
Kim K, Kang DI, Yang J-E (2005) On the use of the balancing method for calculating component RAW involving CCFs in SSC categorization. Reliab Eng System Saf 87:233–242
Kotes P, Vican J (2012) Reliability levels for existing bridges evaluation according to Eurocode. Proc Eng 40:211–2016
Krystek R, Zukowska R (2005) Time series—the tool for traffic safety analysis. In: Kolowrocki (ed) Advances in safety and reliability, Taylor and Francis, London, pp 1199–1202
LeBeau KH, Wadia-Fascett SJ (2007) Fault Tree analysis of Schoharie Creek bridge collapse. J Perform Constr Facil vol 21(4), Aug 2007
Lemaire M, Pendola M (2006) PHIMECA-soft. Struct Saf 28:130–149
Lin H-Z, Khalessi MR (2006) General outlook of UNIPASS V5.0: a general-purpose probabilistic software system. Struct Saf 28:196–216
Liu X (2000) Analysis and control of human errors during structural construction. In: Coble RJ, Haupt TC, Hinze J (ed) The management of construction safety and health. Balkema, Rotterdam, pp 47–66
Low BK, Teh CI (2000) Probabilistic analysis of pile deflection under lateral loads. In: Melchers RE, Stewart MG (eds) Application of statistics and probability ICASP 8, vol 1. Balkema, Rotterdam pp 407–414
Maag T (2004) Risikobasierte Beurteilung der Personensicherheit von Wohnbauten im Brandfall unter Verwendung von Bayes'schen Netzen, Institut für Baustatik, und Konstruktion, ETH Zürich, vdf Hochschulverlag AG an der ETH Zürich, IBK Bericht 282, März 2004 Zürich
MacDonald AJ, Wight RG, Bartlett FM (2016) Acceptable risk in military bridge evaluation. Adv Mil Technol 1(2):197–209
Madsen HO, Krenk S, Lind NC (2006) Methods of structural safety. Dover Publications, Reprint, Mineola/New York
Mainpost (2009) Bei Schiffskollision nicht alle standfest. Main-Netz 6(2):2009
Mannan S (2005) Lee´s loss prevention in the process industries, hazard identification, assessment and control, vol 1, 3rd edn. Elsevier Butterworth-Heinemann, Burlington
Mathieu H, Saillard Y (1974) Sécurité des Structures Concepts générauxh charges et actions. CEB, Bulletin d'information 102, Paris
Matousek M (1982) Massnahmen gegen Fehler im Bauprozess. Birkhäuser: Zürich
Matousek M, Schneider J (1976) Untersuchungen zur Struktur des Sicherheitsproblems bei Bauwerken, IBK-Bericht No, 59, ETH Zürich
Mayer M (1926) Die Sicherheit der Bauwerke und ihre Berechnung nach Grenzkräften anstatt nach zulässigen Spannungen. Verlag von Julius Springer, Berlin
Melchers RE (1984) Human error in structural reliability assessments. Reliab Eng 7:61–75
Melchers RE (1999) Structural Reliability—analysis and prediction, 2nd edn. Wiley, Chichester
Melchers RE, Baker MJ, Moses F (1983) Evaluation of experience. Proceedings—IABSE Workshop 1983: Quality Assurance within the Building Process, Int Assoc Bridge Struct Eng, Rigi, 21–38
Menzies JB (1996) Bridge failures, hazards and societal risk. In: International symposium on the Safety of bridges, July 1996, London
Moses F (2001) Calibration of load factors for LRFR bridge evaluation, NCHRP Report 405, Transportation Research Board. National Research Council, Washington DC

Müller HS, Vogel M, Neumann T (2011) Die Quantifizierung der Lebensdauer von Betonbrücken mit den Methoden der Systemanalyse, Berichte der Bundesanstalt für Straßenwesen, Brücken- und Ingenieurbau, Heft B81, Bergisch Gladbach

Murzewski J (1974) Sicherheit der Baukonstruktionen. VEB Verlag für Bauwesen, Berlin, DDR

O'Connor C, Shaw P (2000) Bridge loads: an international perspective. SPON Press, London

Pan Y, Agrawal AK, Ghosn M (2007) Seismic fragility of continuous steel highway bridges in New York State. J Bridge Eng 12(6):689–699

Petschacher M (1994) VaP a tool for practicing engineers. In: Schueller et al (eds) Proceedings of ICOSSAR'936th international conference on Structural safety and reliability in Innsbruck, Balkema, pp 1817–1823

Proske D (2009) Catalogue of Risk. Springer, Berlin, Heidelberg

Proske D (2012) Vollprobabilistische Ermittlung der Fragility-Kurve einer Stahldruckschale bei Wasserstoff-Deflagration, Bautechnik, vol 89, Heft 1

Proske D, Loos S (2009) Eurocode road traffic load models for weight restricted bridges, ESREL 2009. In: Briš, Guedes Soares, Martorell (eds) Reliability, risk and safety: theory and applications. Taylor & Francis Group, London, ISBN 978-0-415-55509-8, pp 1397–1404

Proske D, van Gelder P (2009) Safety of historical stone arch bridges. Springer, Berlin, Heidelberg

Pugsley A (1968) The safety of bridges. Struct Eng 46(1968):197

Pukl R, Novak D, Vorechovsky M, Bergmeister K (2006) Uncertainties of material properties in nonlinear computer simulation. In: Dirk Proske, Milad Mehdianpour und Lucjan Gucma (eds) 4th International probabilistic symposium, Berlin, pp 127–138

Quan Q (2004) Modeling current traffic load, safety evaluation and SFE analysis of four bridges in China. In: Brebbia CA (ed) Risk analysis IV. Wessex Institute of Technology Press, Southampton

Rackwitz R (1998) Zuverlässigkeit und Lasten im konstruktiven Ingenieurbau. Zuverlässigkeitstheoretische Grundlagen, Technische Universität München, Teil I, pp 1993–1998

Rackwitz R (1999) Zuverlässigkeitsbetrachtungen bei Verlust der Dauerhaftigkeit von Bauteilen und Bauwerken; Kurzberichte aus der Bauforschung 40 (1999), Nr. 4, Forschungsbericht Nr. T 2847, Stuttgart IRB, pp 297–301

Rackwitz R (2001) Zuverlässigkeit von Tragwerken, in: Zilch, Diederichs, Katzenbach (eds) Handbuch für Bauingenieure, Springer, Berlin, pp 1-217-1-260

Randsaeter A (2000) Risk Assessment in the offshore industry, In: Proceedings—Part 2/2 of Promotion of technical harmonization on risk-based decision-making, Workshop, May 2000. Stresa, Italy

Reason J (1990) The Contribution of latent human failures to the breakdown of complex systems. Philos Trans R Soc Lond B Biol Sci 327(1241):475–484

Reh S, Beley J-D, Mukherjee S, Khor EH (2006) Probabilistic finite element analysis using ANSYS. Struct Saf 28:17–43

Reid SG (2000) Acceptable risk criteria. Prog Struct Mat Eng 2(2):254–262

Rizkallah V, Harder H, Jebe P, Vogel J (1990) Bauschäden im Spezialtiefbau, Institut für Bauschadensforschung e, V, Heft 3. Eigenverlag, Hannover

SBA (2006) Statistisches Bundesamt: Unfallgeschehen im Straßenverkehr 2005. Statistisches Bundesamt Wiesbaden, Deutschland

Schlegel R, Will J (2007) Sensitivitätsanalyse und Parameteridentifikation von bestehenden Mauerwerkstrukturen. Mauerwerk 11. Heft 6:349–355

Schneider R, Fischer J, Bügler M, Nowak M, Thöns S, Borrmann A, Straub D (2015) Assessing and updating the reliability of concrete bridges subjected to spatial deterioration—principles and software implementation. Structural Concrete 16(3):356–365

Schueller GI, Pradlwarter HJ (2006) Computational stochastic structural analysis (COSSAN)—a software tool. Struct Saf 28:68–82

Schueremans L, Van Gemert D (2001) Assessment of existing masonry structures using probabilistic methods—state of the art and new approaches. Fifth international symposium on computer methods in structural masonry STRUMAS, KU Leuven, Belgium, 18–20 April 2001

SIA 269 (2007) Schweizerischer Ingenieur- und Architektenverein: SIA 269: Grundlagen der Erhaltung von Tragwerken; Schweizerischer Ingenieur- und Architektenverein, Zürich Entwurf 03/2007
Spaethe G (1992) Die Sicherheit tragender Baukonstruktionen, 2, Neubearbeitete edn. Springer Verlag, Wien
Steel RD, Torrie JH (1991) Prinsip dan Prosedur Statistika, 2nd edn. Gramedia Pustaka Utama, Jakarta
Steenbergen RDJM, Vrouwenvelder ACWM (2010) Safety philosophy for existing structures and partial factors for traffic in loads on bridges. Heron 55(2)
Stewart MG (1993) Structural reliability and error control in reinforced concrete design and construction. Struct Saf 12:277–292
Stewart MG, Melchers RE (1989) Error control in member design. Struct Saf 6:11–24
Straub D, Der Kiureghian A (2011) Reliability acceptance criteria for deteriorating elements of structural systems. J Struct Eng Trans. ASCE 137(12):1573–1582
Strauss A, Bergmeister K (2005) Safety concepts of new and existing structures. Presentation at fib commission 2nd meeting on Safety and performance concepts, Budapest, 23 May 2005
Sýkora M, Holicky M (2013) Target reliability levels for the assessment of existing structures—case study. In: Strauss A, Frangopol DM, Bergmeister K (eds) Life-cycle and sustainability of civil infrastructure systems. CRC Press, Taylor and Francis, London, pp 189
Sýkora M, Holický M, Lenner R, Manas P (2013) Optimum target reliability for bridges considering emergency situations. In: Proceedings of the 11th international probabilistic workshop on Novák and Vořechovský, Brno, pp 439–450
Tanner P, Hingorani R (2015) Acceptable risks to persons associated with building structures. Struct Concr 16(3):314–322
Tarkov J (1986) A disaster in the making. Am Herit Invent Technol 1(3)
Thacker BH, Riha DS, Fitch SHK, Huyse LJ, Pleming JB (2006) Probabilistic engineering analysis using the NESSUS software. Struct Saf 28:83–107
Thoft-Christensen P, Baker M (1982) Structural reliability theory and its applications, Murotsu, Yoshisada: application of structural system reliability theory. Springer, Berlin, Heidelberg, New York
Tichý M (1976) Probleme der Zuverlässigkeit in der Theorie von Tragwerken, Vorträge zum Problemseminar: Zuverlässigkeit tragender Konstruktionen, Weiterbildungszentrum Festkörpermechanik, Konstruktion und rationeller Werkstoffeinsatz, Technische Universität Dresden—Sektion Grundlagen des Maschinenwesens, Heft 3/76
Tsang HH, Wenzel F (2016) Setting structural safety requirements for controlling earthquake mortality risk. Saf Sci 86: 174–183
Tvedt L (2006) Proban—probabilistic analysis. Struct Saf 28:150–163
Tweed MH (1969) A summary and analysis of bridge failures. M.Sc. Thesis, Iowa State University
van der Borst M, Schoonakker H (2001) An overview of PSA importance measures. Reliab Eng Syst Saf 72:241–245
Vogel T, Zwicky D, Joray D, Diggelmann M, Hoj NP (2009) Tragsicherheit der bestehenden Kunstbauten, Sicherheit des Verkehrssystems Strasse und dessen Kunstbauten, Bundesamt für Strassen, Dezember 2009, Bern
Voigt J (2014) Beitrag zur Bestimmung der Tragfähigkeit bestehender Stahlbetonkonstruktionen auf Grundlage der Systemzuverlässigkeit. Dissertation, Siegen
Vrijling JK, van Gelder PHAJM, Ouwerkerk SJ (2005) Criteria for acceptable risk in the Netherlands, Infrastructure Risk Management Processes, pp 143–157
Wang N (2010a) Reliability-based condition assessment of existing highway bridges. Dissertation, Georgia Institute of Technology, Aug 2010
Wang N (2010b) Reliability-based condition assessment of existing highway bridges. Dissertation, Georgia Institute of Technology, Aug 2010

Weber M, Caspeele R, Schnell J, Glock C, Botte W (2018) Das neue fib Bulletin 80—Teilsicherheitsbeiwerte für die Nachrechnung bestehender Massivbauwerke. Beton- und Stahlbetonbau (in print)

Wu Y-T, Shin Y, Sues RH, Cesare MA (2006) Probabilistic function evaluation system (ProFES) for reliability-based design. Struct Saf 28:164–195

Zentner I, Nadjarian A, Humbert N, Viallet E (2008) Estimation of fragility curves for seismic probabilistic risk assessment by means of numerical experiments. In: Graubner C-A, Schmidt H, Proske D (eds) 6th international probabilistic workshop, Darmstadt, Germany 2008, Technische Universität Darmstadt, pp 305–316, 26–27 Nov 2008

Zerna W (1983) Grundlage der gegenwärtigen Sicherheitspraxis in der Bautechnik. In: S. Hartwig (ed) Große technische Gefahrenpotentiale: Risikoanalysen und Sicherheitsfragen, Hrsg. Springer Verlag: Berlin/Heidelberg, pp 99–109

Chapter 6
Collapse Frequencies of Bridges

6.1 Introduction

As mentioned in the Chap. 3, either the direct bridge collapse frequency numbers or the data used to determine the bridge collapse frequency within this book are taken from various publications and scientific works. Hence no new data is added, however the existing data is combined. The extended data pool allows more robust conclusions and perhaps some new conclusions in comparison to the conclusions drawn in each individual study.

6.2 Data Basis

In this section the studies and databases used are listed. However, not all studies could be used for all types of analysis or in the same manner. For example, terms such as "bridges" or "causes of bridge collapses" have been defined differently in different studies. Therefore different studies and different parts of studies respectively have been selected for different parts of the analysis carried out within this book.

All studies known are listed in Table 6.1. The studies are sorted alphabetically according to the first author's name. Some of the studies use data from other studies, then the studies are either put together in the list or it is mentioned. Some studies do not cover the entire bridge population, e.g. therefore they use censored data. Often the censoring criterion is the availability of information in the public media.

In some studies the limitations are obvious. For example, McLinn (2010) states that 40 bridges per year collapsed in the U.S. before 1900. There are no further details. Dubbudu (2016) gives only the number of fatalities due to bridge collapses in India, not the number of bridge collapses. However, all these data is included in the table for the sake of completeness. There may be more such databases worldwide;

D. Proske, *Bridge Collapse Frequencies versus Failure Probabilities*, Risk Engineering,
https://doi.org/10.1007/978-3-319-73833-8_6

Table 6.1 Table of studies

References	Region	Time period	Number of bridge collapses
Breysse and Ndiaye (2014)	World	n/a	n/a
Bridge Forum (2017)	World	1444–2009	360
Brückenweb (2017): Accidents and collapses	World	1157–2017	284
Christian (2010)/Briaud et al. (2012)	U.S.	1966–2005	1502
Cook (2014)	U.S.	1987–2011	103
Diaz et al. (2009)	Colombia	1986–2008	63
Dubbudu (2016): only fatalities given (297)	India	2010–2014	n/a
Fard (2012)		1818–2012	100
Fu et al. (2012)	China	2000–2012	157
Harik et al. (1990)	U.S.	1951–1988	77
Hersi (2009)	U.S.	2000–2008	161
Imam and Chryssanthopoulos (2012): Metallic bridges	World	Early 19th century–2011	164
Imhof (2004)	World	1444–2004	347
Lee et al. (2013a)	World	1980–2012	1062
Lee et al. (2013b)	World	1876–2005	1723
McLinn (2010)	U.S.	Before 1900	40 per year
McLinn (2010): Major collapses only	World	1970–2009	71
Menzies (1996)	U.K.	n/a	n/a
Montalvo and Cook (2017)	U.S.	1992–2014	428
Scheer (2010)	World	813–2008	536
Sharma (2010)	World	1800–2009	1814
Sharma and Mohan (2011)	U.S.	1800–2009	1367
Smith (1976)	U.S.	1847–1975	143
Taricska (2014)	U.S.	2000–2012	341
Vogel et al. (2009)	Switzerland	Mainly Imhof data	
Vogel et al. (2009)	World	Mainly Imhof data	
Wardhana and Hadipriono (2003)	U.S.	1989–2000	503
Wikipedia (2017)	World	1297–2017	242
Xu et al. (2016)	China	2000–2014	302
Zerna (1983): Steel bridges	U.S.	Before 1900	1
Zerna (1983): Suspension bridges	World	1900–1940	7

however they are not accessible by the author. Finally, care has to be taken to avoid, that the limitations of the individual studies affect the results of the meta-analysis.

Table 6.2 lists the databases behind the individual studies. This table is of utmost importance because if all studies depend on one database they are not independent and results can not be considered as additional confirmation in terms of a meta-analysis. The table shows, that many studies from the U.S. are somehow related to each other since they use, at least partly, the same database. The studies can be seen rather like a re-sampling. However, this fact is to a certain extend not surprising, since the number of samples drawn (collapses) and the information regarding the samples is limited.

Table 6.3 lists the parts of information used within this book to draw conclusions.

6.3 Number of Bridges Worldwide

According to Chap. 2, the number of the reference population is necessary to compute the collision frequency. This is done first for the global inventory.

Bridge structures are an essential part of the infrastructure system worldwide. They significantly contribute to the serviceability and the maintenance of modern human societies which are characterised by large information-, energy- and material flows. In Germany, weight and volume of road traffic have grown exponentially in the last decades (Naumann 2002). Hannawald et al. (2003) measured a 70 tonne truck on German highways under regular traffic conditions and Pircher et al. (2009) and Enright et al. (2011) report the measurement of 100 tonne trucks. Further extensions of weight and dimensions of heavy commercial vehicles are planned by the European Commission (Directive 96/53/EC).

The high traffic needs in industrialised countries yield to about one bridge per 500 inhabitants and in developing countries to about one bridge per 2000 inhabitants. Based on these estimations the worldwide number of bridges is between five and six million. These numbers are in compliance with Vogel et al. (2009) who estimate five million bridges.

Of course, the estimation of the number of bridges based on geographic conditions (topography, population density, industry etc.) would provide a better estimate. For example Weber (1999) gives estimates of bridge densities on railway lines for certain countries with large ratios of Alpine region to the overall country area. For example, the Swiss railway SBB reaches a bridge density value of 14 (number of bridges per 10 km of operated railway distance) in comparison to the Polish and Danish Railways, which reach density values of 4 and 6 respectively (Weber 1999).

However, such data is not available on a large scale. Also public accessible databases such as the one for the U.S. bridge stock (FHWA 2017) or partly for the German bridge stock (Nagel et al. 2016) are not available worldwide. Therefore mainly the data in Table 6.4 is used for the determination of the bridge stock. This data is also illustrated in Fig. 6.1.

Table 6.2 Databases used (abbreviations see below)

References	Sources	Independent
Breysse and Ndiaye (2014)	Unknown	Unknown
Bridge Forum (2017)	Journals, News media, books	Independent
Brückenweb (2017): Accidents and collapses	Journals, News media, books	Independent
Christian (2010)/Briaud et al. (2012)	Data from Briaud, details unknown	
Cook (2014)	NYDOT database, Regional bridge failure database (DOT questionnaire)	Partly independent
Diaz et al. (2009)		Independent
Dubbudu (2016): only fatalities given (297)	Indian statistics	Independent
Fard (2012)		
Fu et al. (2012)	Own database	Independent
Harik et al. (1990)	Unknown	
Hersi (2009)	NYDOT database, Home Pages, FHWA, ASCE Journal Publications, further Journal Publications	Partly independent
Imam and Chryssanthopoulos (2012): Metallic bridges	Literature, Web, News reports	Independent
Imhof (2004)	Own database	Independent, related to Menzies (?)
Lee et al. (2013a)	NYDOT database, many journals, Scheer (2010), AASHTO reports, MCEER reports, News reports	Partly independent
Lee et al. (2013b)	Integrates current exiting bridge information—see Lee et al. (2013a)	Partly independent
McLinn (2010)	Own database	Independent, but limited
McLinn (2010): Major collapses only	Own database	Independent, but limited
Menzies (1996)	Das (1997) states, that the collapse data base may be limited (too low)	Independent
Montalvo and Cook (2017)	NYDOT, NBI	Partly independent
Scheer (2010)	Journals, News media	Independent
Sharma (2010)	–	–
Sharma and Mohan (2011)	Own database, details unknown	Independent
Smith (1976)	Own database, details unknown	Independent
Taricska (2014)	NYDOT database, journals, home pages, FHWA, ASCE Journal publications, further Journal publications	Partly independent

(continued)

Table 6.2 (continued)

References	Sources	Independent
Vogel et al. (2009), Switzerland	Some data from Imhof (2004) and data added	Partly independent from Imhof (2004)
Vogel et al. (2009), Worldwide	Mainly data from Imhof (2004), data added	Minor independent from Imhof (2004)
Wardhana and Hadipriono (2003)	NYDOT database, journals, home pages, FHWA, several departments of transportation NY, Ohio, Utah, Wisconsin, Texas, Illinois, personal experience, E-mails	Partly independent
Wikipedia (2017)	Journals, News media, books	Independent
Xu et al. (2016)	Own database	Independent
Zerna (1983): Steel bridges	Unknown	Independent
Zerna (1983): Suspension bridges	Unknown	Independent

AASHTO American Association of State Highway and Transportation Officials
ASCE American Society of Civil Engineers
DOT Department of Transportation
FHWA Federal Highway Administration
MCEER Multidisciplinary Center for Earthquake Engineering Research
NY New York
NBI National Bridge Inventory
NYDOT New York Department of Transportation

If one assumes, that all countries worldwide reach the number of bridges related to the population in developed countries and assuming a world population of 7.5 billion people it would yield to a number of 15 million bridges. However, the human population is expected to growth up to 10 billion in the year 2050. This would even yield to an upper estimate of 20 million bridges on earth in 2050.

The current population in developed countries is in the order of 1.5 billion people. That alone yields to 3 million bridges. Assuming that the population in these countries will not significantly grow the number of bridges in these countries will also not significantly grow in the next decades.

Based on the figures from China in recent decades the number of bridges nearly doubles every ten years and thus shows a large growth of the bridge stock. If we generalise this grow and assume that countries with 50% of the world's population will transform from developing countries to developed countries and therefore change the bridge-to-population-ratio from 1:2000 to 1:500 we will see in the next decades bridge numbers of 2 million in China and India alone.

In other terms the major part of the bridge stock worldwide will not be anymore in the U.S. and Europe, but in Asia, South America and Africa. However this new part of the bridge stock does not need to repeat the failures done the centuries ago in Europe. If we learn appropriately from historical disasters we do not have to see the same bridge collapses we have observed in the decades ago and recently (Xu et al. 2016).

Table 6.3 Data used within the book

References	Cause	Age	Material	Static system	Further remarks
Breysse and Ndiaye (2014)					
Bridge Forum (2017)					
Brückenweb (2017): Accidents and collapses					
Christian (2010)/Briaud et al. (2012)					Statistics
Cook (2014)	Yes	Yes	Yes		Statistics
Dubbudu (2016): only fatalities given (297)					
Fard (2012)					
Fu et al. (2012)					
Harik et al. (1990)	Yes				Statistics
Hersi (2009)	Yes				Statistics
Imam and Chryssanthopoulos (2012): Metallic bridges	Yes	Yes	Yes	Yes	
Imhof (2004)	Yes				Statistics
Lee et al. (2013a)	Yes	Yes	Yes		Statistics
Lee et al. (2013b)					
McLinn (2010)					
McLinn (2010): Major collapses only					
Menzies (1996)					Statistics
Montalvo and Cook (2017)					
Scheer (2010)	Yes				
Sharma (2010)					
Sharma and Mohan (2011)				Yes	Statistics
Smith (1976)					
Taricska (2014)	Yes	Yes	Yes	Yes	Statistics
Vogel et al. (2009), Switzerland					Statistics
Vogel et al. (2009), World					Statistics
Wardhana and Hadipriono (2003)	Yes		Yes	Yes	Statistics
Wikipedia (2017)					
Xu et al. (2016)					
Zerna (1983): Steel bridges			Yes		Statistics
Zerna (1983): Suspension bridges			Yes		Statistics

Table 6.4 Bridge Stock in selected countries

Country	Population in Millions	Number of bridges	References
U.S.		10,000	[a]1900
U.S.	260	600,000	Dunker (1993)
U.S.	319	607,380	ASCE (2013)
U.S.		614,387	ASCE (2017)
Europe	743	500,000[b]	Casas (2015)
Europe	743	1,000,000[c]	Casas (2015)
U.K.	58	100,000	Menzies (1996)
U.K.	59	150,000	Woodward et al. (1999)
Germany	81	120,000	Anonymus (2004), Naumann (2004)
Germany	82	120,000	Own estimate[d]
Japan	127	155,000[e]	MLIT (2005)
China	1325	500,000	Yan and Shao (2008)
China	1344	689,400	Zhang et al. (2014)
China	1371	750,000	Xu et al. (2016)

[a]Own estimate
[b]Only road bridges
[c]Only railway bridges
[d]Own estimate: Highway bridges: 39,535 (2017), Railway 25,000 (2017), The city Düsseldorf has 390 non-highway bridges (Vollrath and Tathoff 2002) and 635,000 inhabitants (2017). This yields to a ratio of one bridge per 1628 inhabitants. By applying this number to the overall population of 81 million one receives another 50,000 bridges. The three numbers add up to approximately 115,000 bridges. Assuming a one percent growth in the number of bridges per year, today Düsseldorf has around 430 bridges. This gives approximately 55,000 non-highway bridges and 120,000 bridges overall
[e]Only road bridges longer than 15 m

6.4 Collapse Frequency of Bridges

As mentioned in Chap. 5, structures in general and bridges specifically can be designed based on probabilities of failure. However, this probability of failure is usually not considered to be comparable to the collapse frequency. Many possible influence factors, such as human errors or correlations between different limit states seem neither to be considered at all nor considered to an adequate extent. This understanding is even explicitly included in codes, such as Eurocode 0 (2017), Table 2.3, Note 2. Spaethe (1992) states that "the probability of failure is only a part of the overall collapse frequency … further contributions such as human failures are not included in the theoretical value. If one assumes an error free mechanical model then the collapse frequency is expected to be larger then the probability of failure." Also, Vogel et al. (2009) refer to the limited comparability of the

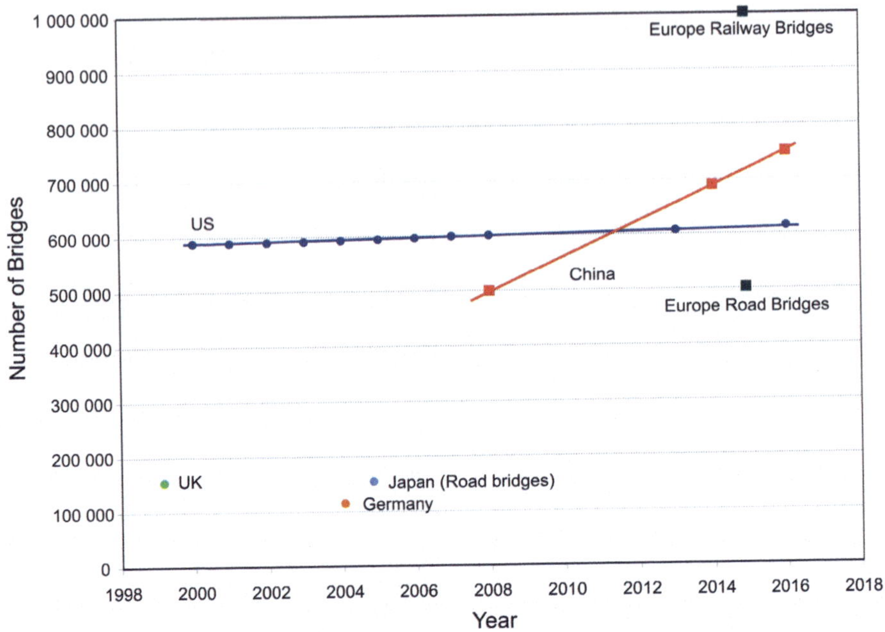

Fig. 6.1 Illustration of the bridge stock for different countries according to Table 6.2

probability of failure and the collapse frequency. Ellingwood (2001) concludes that "the probability estimates may not correspond to historical failure rates."

On the other hand, various uncertainties are used reasoning the application of probabilistic computations such as in loads, material properties, dimensions, natural and manmade hazards, insufficient knowledge and human errors in design and construction (Ellingwood et al. 1980, 1982, Wang 2010). However, it seems illogical to use human failure as reason for the application of probabilistic methods and then to exclude it from the computation. Based on this contradiction we dare to compare the probabilities of failure of bridges and the collapse frequencies in this book and should be able to quantify the influence of human failure.

Brown (1979) stated that the computed probability of failure is usually 10^{-6} per year whereas the observed collapse frequencies seemed to lie between 10^{-2} and 10^{-3} per year. This statement has been widely referred to such as Elm (1998), Nendo and Niczyj (1998). In contrast, Nowak and Collins (2012) give collapse frequencies between 10^{-3} and 10^{-5} per year. If the observation by Brown (1979) is true, it would show an extreme difference between the probabilities of failure and the collapse frequencies. If the probability of failure is supposed to be a true (systematic error free), robust, converging, and useful parameter it has to have a relation to observations such as the collapse frequency. Even further, the codes such as Eurocode often permit both techniques equally for the determination of design values, the probability of failure and real world observations.

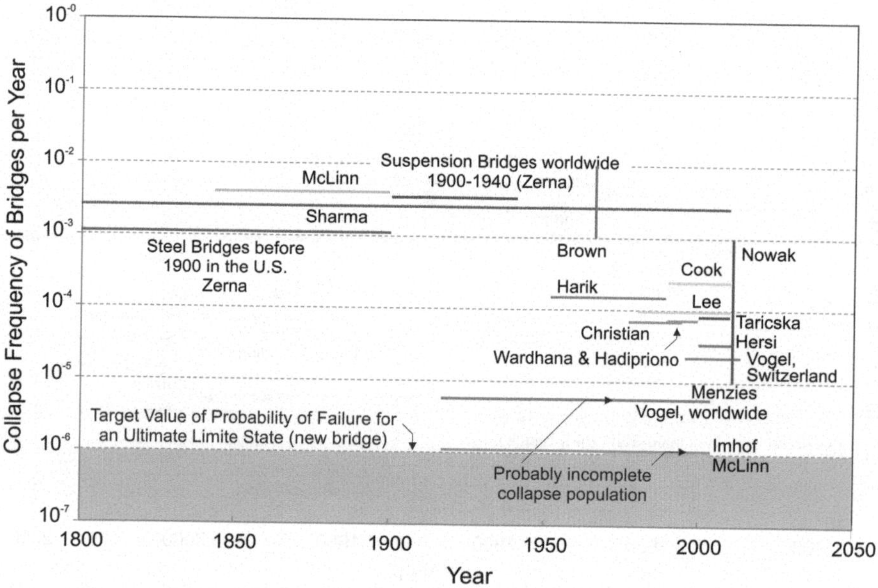

Fig. 6.2 Collapse frequency of bridges per year based on various publications (time periods)

Figure 6.2 shows the bridge collapse frequencies given by Brown (1979), Nowak and Collins (2012) as vertical lines and most of the studies listed in Table 6.1 as horizontal lines over the time period. Although Browns (1979) study is related to structures in general, it complies with bridges of the 19th century. Maximum numbers reach 4×10^{-3} per year based on the number of bridge collapses given by McLinn (2010) and own estimations of the bridge stock. This number is confirmed by Zerna (1983), Sharma and Mohan (2011). So, one can summarize that the overall bridge collapse frequency in the U.S. before 1900 was approximately 10^{-3} per year and bridge.

However looking at the data after 1950, the collapse frequencies indicate a significant improvement. All studies show a collapse frequency below 10^{-3} per year, most of the studies give a collapse frequency below 10^{-4} per year.

Even further, some studies show surprisingly low collapse frequencies compared to the majority of studies, such as the values by Imhof (2004), Vogel et al. (2009) for worldwide data, Menzies (1996) and McLinn (2010). These numbers are probably based on incomplete populations. This fact has at least been confirmed by Imhof in a personal communication to the author. Since Vogel et al. (2009) worldwide collapse data is based on Imhof (2004) data the conclusions is also true for this publication although Vogel et al. (2009) have partly considered this limitation.

Figure 6.3 shows the same data as Fig. 6.2 but introducing an anchorage or centre point of the time period covered by the data. The selection of the location of this point will be discussed later, however it can be seen that the data considering the time after 1950 forms a cluster. This cluster is bordered at the left by Harik (1990)

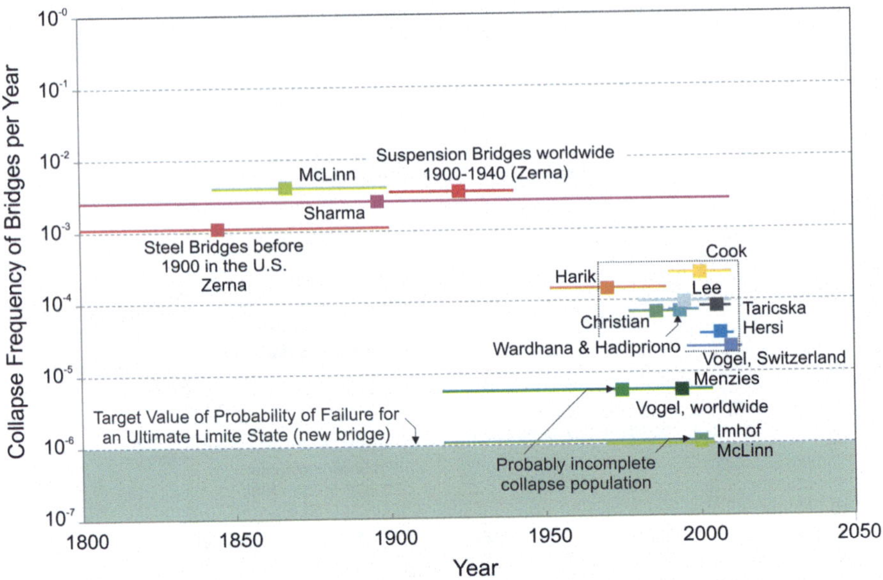

Fig. 6.3 Collapse frequency of bridges per year based on various publications (time periods)

and on the right by Vogel et al. (2009), it is bordered at top by Cook (2014) and at the bottom by Vogel et al. (2009). The cluster is framed in Fig. 6.3 by an rectangle. The data points indicate an average collapse frequency in the range of 10^{-4} per year.

To compare the importance of individual data points, Fig. 6.4 shows the sample sizes as area of a circle. The figure clearly shows that some references include very large sample sizes whereas others include only a small number of collapses. On one hand, this limits the value of the data points, for example from Zerna (1983) or from Vogel et al. (2009) for Switzerland. On the other hand, Switzerland may have shown such an overall small number. The figure also confirms that the sample size of the very low collapse frequencies is too low, because the global data (Imhof 2004; Vogel et al. 2009 worldwide) are smaller than country-specific samples (Lee et al. 2013a), which is not possible by definition.

In general, all data points in the diagram are relatively independent using the conclusion from Table 6.2. However, most data points inside the rectangle are related since they partly use the same databases such as Cook (2014), Lee et al. (2013a, b), Taricska (2014), Wardhana and Hadipriono (2003). This has the advantage that they use all the same bridge definition (length of span), it has the disadvantage that the data points can not be considered as fully independent samples. Lee et al. uses data from Scheer (2010) which may probably use another bridge definition (length of span) but will increase the independence of the data points. The data in the cluster is related to industrialized countries and recent decades. The cause of the deviation of the data points will be discussed in the section cause of collapses.

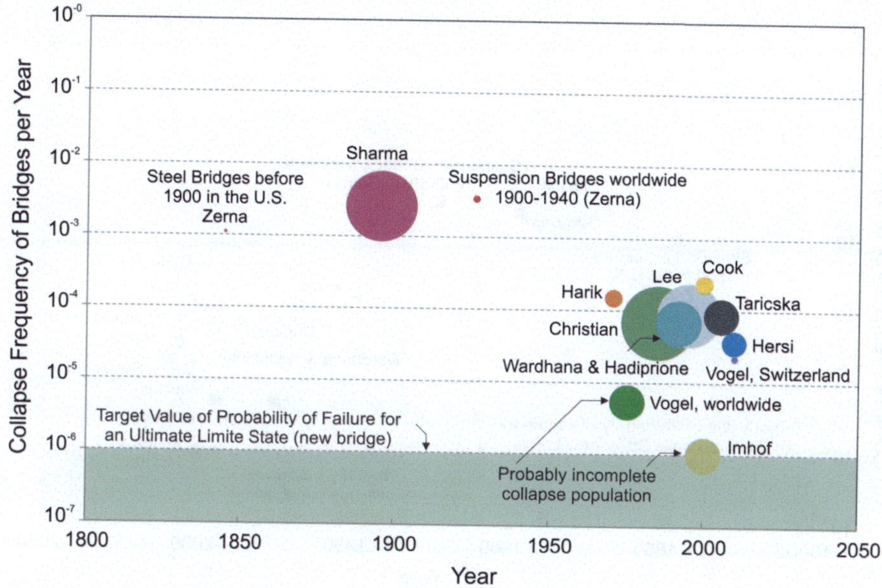

Fig. 6.4 Collapse frequency of bridges per year considering the number of collapses per reference as size of the circle

6.5 Time-Dependency

Figures 6.2, 6.3, 6.4 and 6.5 combine the collapse frequencies of bridges given by various authors for different time periods and different regions (for references see Table 6.1). As can be clearly seen there is a decreasing trend over the last century. The collapse frequency of bridges has decreased by nearly two orders of magnitude.

In Fig. 6.5 the trend is shown by two lines, a linear decreasing trend given by Cook (2014) and a nonlinear trend introduced by the author. Please note: since the collapse frequency is given in log-scale, actually both functions are nonlinear.

However, the nonlinear trend depends heavily on the collapse frequency numbers given by Zerna (1983), McLinn (2010), Sharma and Mohan (2011). Their numbers are located outside the cluster of collapse frequencies related to modern times and therefore strongly influence the shape of the function.

Since they not only cover very early times but also very long periods (over 200 years) their anchorage points in the diagram can vary over a wide range. For the Figs. 6.3 and 6.4 the middle of the period has been used. However one can argue that the early collapse frequencies are dominating the average collapse frequencies over the long period, since they might change by one or two orders of magnitude. This conclusion would mean that the data points from Sharma and Mohan (2011), McLinn (2010) and less from Zerna shift to the left. Hence the curve would become more linear and the slope would decrease: in other terms the improvement of the

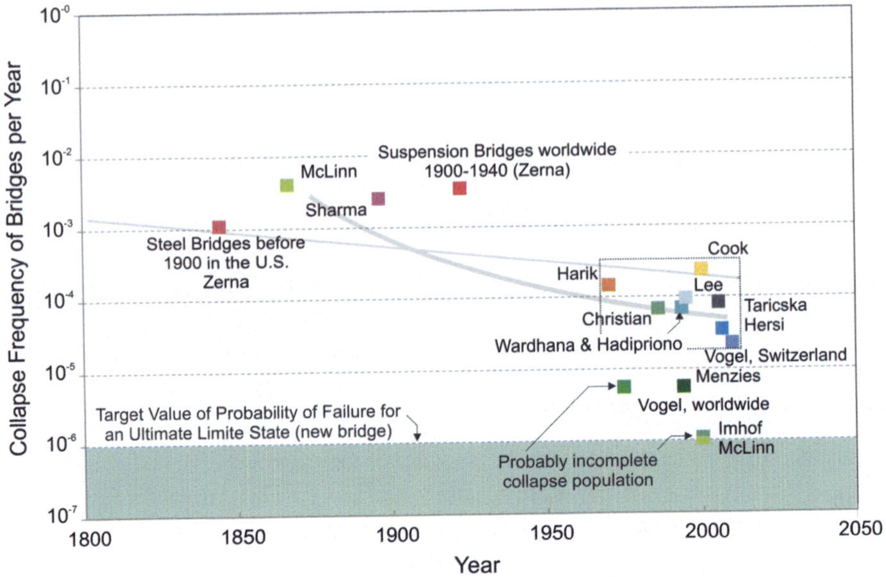

Fig. 6.5 Trend of the collapse frequency of bridges based on various publications

safety is smaller in the last century then expected and it would confirm the curve given by Cook (2014).

A simple regression is given with

$$F_C(t) = 38673 \cdot \exp(-0.0096 \cdot t)$$

with t as calendar year (for example 2014).

Unfortunately the variation between the data points in the last decades is still so high that it masks the time-dependency in the last decades.

6.6 Causes of Damages and Conclusions

6.6.1 Introduction

As mentioned above, the difference in the cluster is still in the range of one order of magnitude. It would be interesting to identify the cause of this difference. Therefore the cause of collapses for the references with high collapse frequencies is compared to the references with low collapse frequencies within the cluster.

Figure 6.6 illustrates the results as histogram of the triggering cause of bridge collapses. Cook (2014) and Harik et al. (1990) show the highest collapse frequencies.

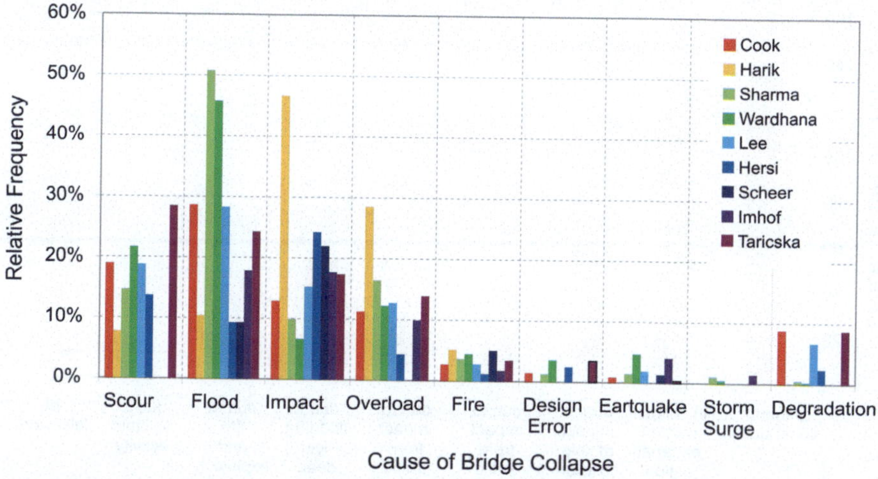

Fig. 6.6 Causes of collapse of bridges based on various publications

In Cook (2014) collapse is dominated by flooding (scour and flood), whereas in Harik et al. (1990) impact and overload are dominating. Since for Vogel et al. (2009) no detailed information is available, Scheer (2010) data has been used, which should be comparable due to the neighbourhood regions (Switzerland and Germany). Based on Scheer (2010) most collapses in Germany are related to the construction time and to impact, see Fig. 6.6. In contrast, Fig. 6.7 indicates a large number of collapses without external actions. Based on a work by Matousek from 1976, Imhof (2004) also found that most of the collapses occur during construction. However, he also refers to other works which do not confirm this fact. The importance of falsework and scaffold failure as shown by Scheer (2010) has also been shown by other references such as Anumba et al. (2006), see Fig. 6.8.

Based on the Fig. 6.6 it can be concluded that the highest collapse frequencies for the data in the rectangle (Fig. 6.3) as given by Cook (2014), Harik et al. (1990) and Lee et al. (2013a, b) for constructed bridges are mainly based on accidental loads. Flooding is also the highest risk in Smith (1976) with almost 50%.

Figure 6.9 shows the average of the bridge collapse causes which is in full compliance with the long measurement series by Lee et al. (2013b).

For the sake of completeness in Figs. 6.10, 6.11 and 6.12 causes for bridge failures are given according to Imhof (2004), Biezma and Schanack (2007). The interpretation of these figures is more difficult than Figs. 6.6, 6.7, 6.8 and 6.9 since enabling and triggering causes are mixed. Still collapses caused by flooding and impact contribute by almost 40%. Therefore the former conclusion remains valid. Only the data by Biezma and Schanack (2007) shows a larger contribution by structural and design deficiencies.

It is well known that accidental loads significantly influence the collapse frequency since the collapse of many bridges in one year due to one major event has strong

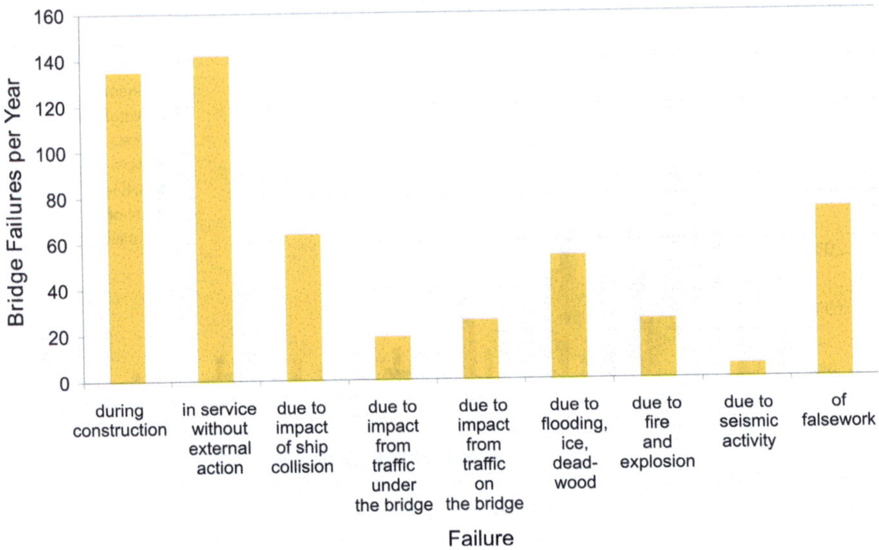

Fig. 6.7 Causes of collapse of bridges based Scheer (2010)

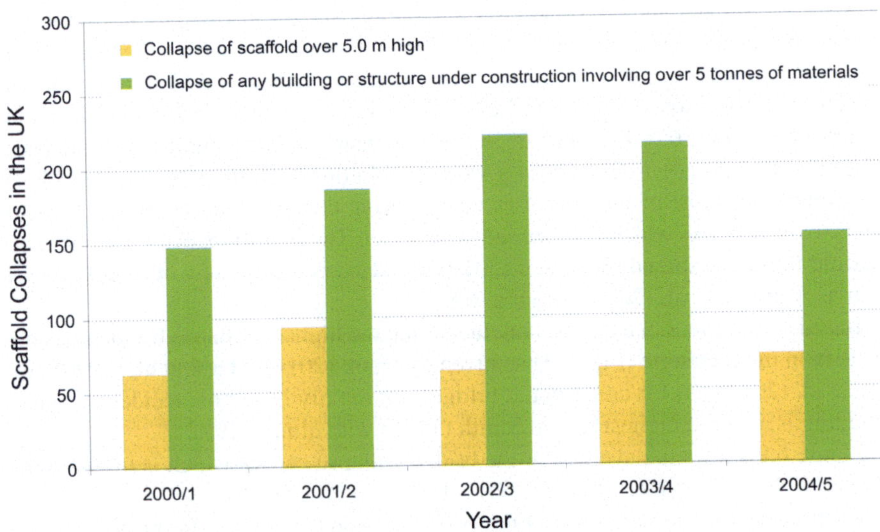

Fig. 6.8 Scaffold collapse and structural collapse during construction related to all types of structures under construction (Anumba et al. 2006)

effects on the collapse number. In the following paragraphs examples of such events are given.

In August 1952 the flood at Lynmuth damaged or destroyed 28 bridges (Hamill 1999). In 1976, typhoon Fran hit Japan and washed 233 bridges away (Hamill 1999).

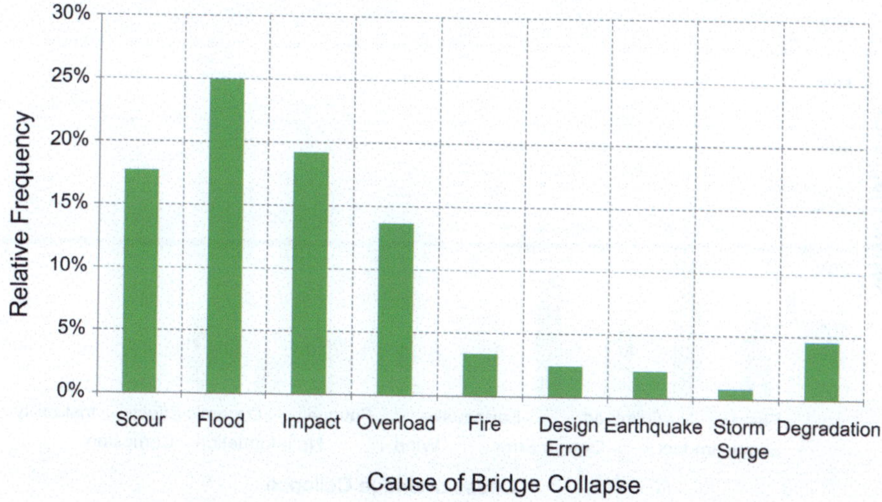

Fig. 6.9 Average of the causes of collapse of bridges based on various publications

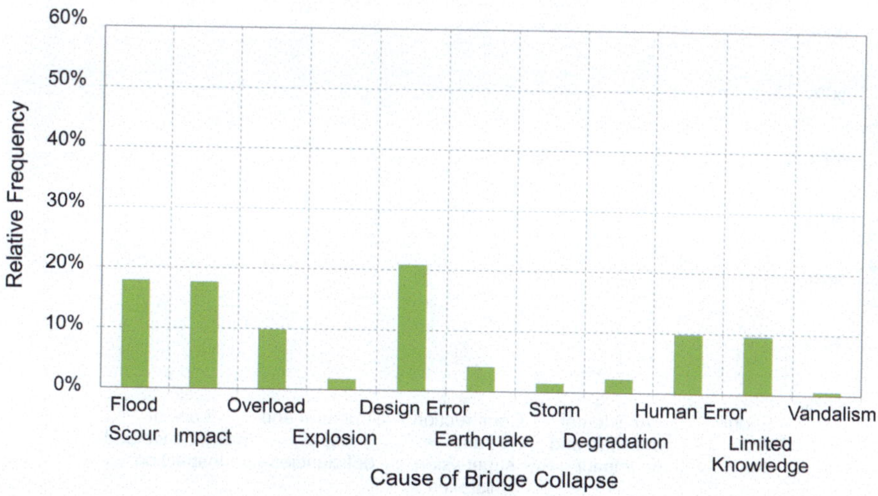

Fig. 6.10 Causes of collapse of bridges based on Imhof (2004)

In 1985 in Pennsylvania, Virginia and West Virgina, 73 bridges were destroyed by flooding (Hamill 1999). In 1987, 17 bridges in New York and New England were damaged or destroyed by spring floods (Hamill 1999). In Germany, several bridges collapsed (Pöppelmannbridge Grimma, Vorlandbridge Riesa—see Reichelt and Richter 2003, Muldebridge Eilenburg) and several hundred bridges were heavily damaged during the flood 2002 (Kraus 2012; Stulc 2015; von Kirchbach et al. 2002; Lehmann 2003, see Figs. 6.13 and 6.14). During Typhoon Sinlaku in September 2008, six bridges collapsed (Hong et al. 2012). Six bridges collapsed in Cumbria,

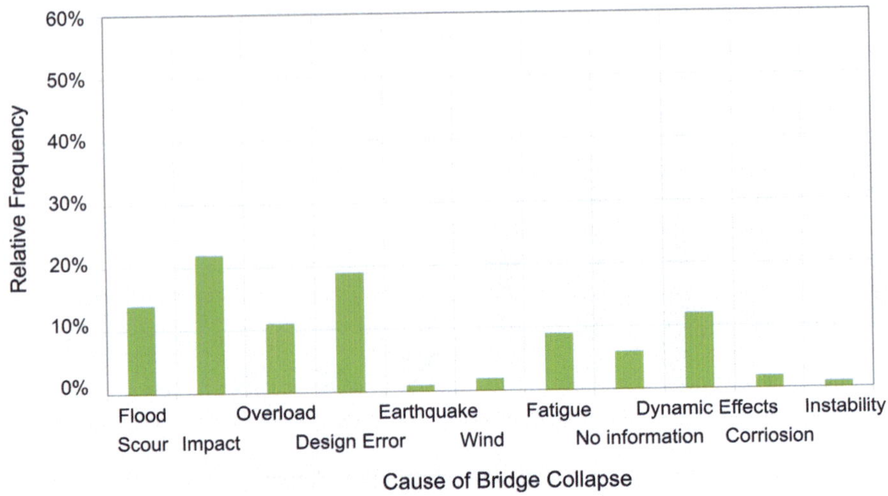

Fig. 6.11 Cause of collapse of bridges based on Bailey taken from Imhof (2004)

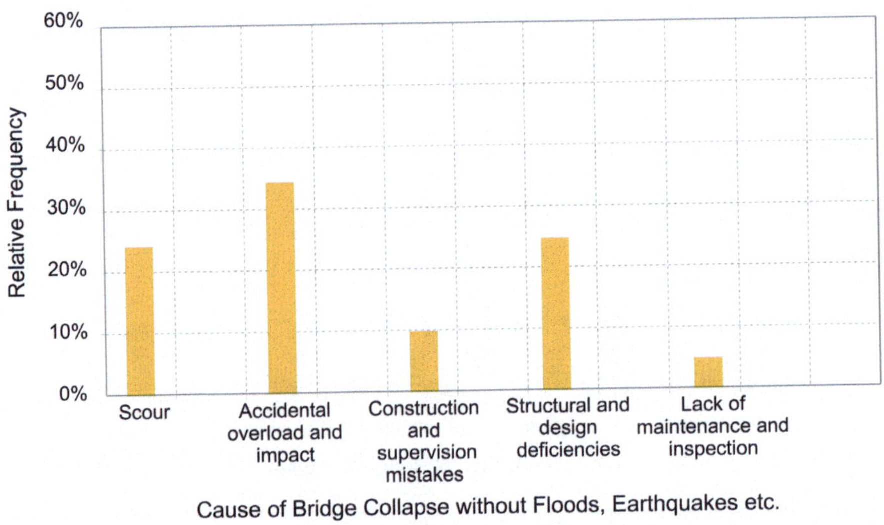

Fig. 6.12 Cause of collapse of bridges based Biezma and Schanack (2007)

U.K. during the November 2009 flood (Cabinet Office 2015). The earthquake and tsunami 2011 in Japan destroyed more than 300 bridges (Maruyama et al. 2012; Akiyama et al. 2012). In spring 2012 in Afghanistan more than 400 small bridges were destroyed by floods (Shorder 2014). Some numbers represent the loss of more than one-tenth of a percent of a countries bridge stock.

Besides hydraulic loads also seismic loads can affect many bridges at once. For example during the San-Fernando-earthquake seven bridges collapsed, during the

Fig. 6.13 Bridge during the 2002 flood in the German city Dresden (*Picture* H. Michler)

Loma-Prieta-earthquakes the Cypress-Street-Viaduct collapsed and killed 41 people. Also several bridges failed during the Kobe-earthquake (Wei et al. 2008; Wenk 2005). The collapse of bridges during large scale disasters is a double loss, not only the direct casualties of the bridge collapse count, also the fact, that the bridge can not be used for emergency and rescue actions. Therefore nowadays modern bridges in regions with high seismicity are designed to deal with higher loads in order to function after a large earthquake.

Other natural hazards only relate to a small population of bridges, for example debris flows. Nevertheless, they can lead to catastrophes and can claim many victims. On July 9th 1981 the pier of the railway bridge above the Liziyida-Ravine on the railway track between Chengdu and Kunmin, China was hit by a debris flow. The pier and consequently the bridge collapsed resulting in the overturning of the train and the death of more than 200 people (Zhang 1993). McSaveny and Davies (2005) described the repeated destruction of a bridge in New Zealand caused by debris flows. The bridge, with a span of 10 m, was destroyed twice in about 11 years. On January 12th–13th 1983 the bridge was destroyed by a debris flow event. The new bridge disappeared after a debris flow in 1994. The bridge was reconstructed to a standard that was described in the media as "bomb-proof". This bridge still exists today. Jan and Chen (2005) described the damage of a bridge at the Chushui River in Shenmu, Taiwan 1996, after a debris flow. In Log Pod Mangartom in Slovenia

Fig. 6.14 Pöppelmann-Bridge in Grimma after the 2002 Flood (*Picture* T. Bösche)

a debris flow destroyed several houses and two bridges in the night of November 16th–17th 2000 (Zorn and Komac 2008). On January 2nd 2012 a suspension bridge was destroyed close to Haselegg, Switzerland by a debris flow (Oberländer 2012). Table 6.5 shows that debris flows contribute with a low one digit percentage value to the collapse frequencies of bridges.

In 1981 a 10 m^3 rock hit the new Gotthard road bridge near Bedrina, Switzerland. The rock nearly missed the pier of the pre-stressed bridge (Bozzolo 1987). Another example is the destruction of a bridge in the Yosemite national park during the Happy Isles rockfall (Morrissey et al. 1999). The listing shows that hazards affecting only a few bridges can nevertheless lead to considerable victims.

6.6.2 Bridge Location

Based on the former considerations we can conclude that the location of the bridge is essential for the estimation of the collapse frequency. For example, all bridge failures due to natural hazards discussed above took place in certain geographic regions showing such hazards. Flashfloods, avalanches, rockfalls, debris flows and

Table 6.5 Detailed type and number of collapse causes

Collapse cause	Wardhana and Hadipriono (2003)		Cook (2014)	
	Number of occurrences	Percentage	Number of occurrences	Percentage
Hydraulic	266	52.88	379	54.85
Flood	165	32.80	198	28.65
Scour	78	15.51	131	18.96
Debris	16	3.18	23	3.33
Drift	2	0.40	2	0.29
Hydraulic	–	–	14	2.03
Ice	–	–	11	1.59
Others	5	0.99	–	–
Collision	59	11.73	89	12.88
Auto/truck	14	2.78	14	8.97
Barge/ship/tanker	10	1.99	11	2.03
Train	3	0.60	2	0.29
Other	32	6.36	–	–
Overload	44	8.75	78	11.29
Deterioration	43	8.55	61	8.83
General	22	4.37	49	1.74
Steel deterioration	14	2.78	12	7.09
Steel corrosion	6	1.19	–	–
Concrete corrosion	1	0.20	–	–
Fire	16	3.18	19	2.75
Construction	13	2.58	10	1.45
Ice	10	1.99	–	–
Earthquake	17	3.38	6	0.87
Fatigue-steel	5	0.99	5	0.72
Design	3	0.60	4	0.58
Soil/Bearing	3	0.60	2	0.29
Sturm/Hurrican/Tornado	2	0.40	17	2.46
Tree fall	–	–	2	0.29
Miscellaneous/other	22	4.37	7	1.01
Total	503	100.00	691	

Table 6.6 Cause-proportional conditional collapse rate for bridges in the U.S. (Cook 2014)

Collapse cause	Over water		Over roadways and railways	
	Number of occurrences	Percentage	Number of occurrences	Percentage
Hydraulic total	379	62.23	7	8.05
Collision total	42	6.90	52	59.77
Overload	69	11.33	3	3.45
Deterioration total	55	9.03	6	6.90
Fire	12	1.97	6	6.90
Construction	7	1.15	4	4.60
Earthquake	3	0.49	2	2.30
Fatigue-steel	4	0.66	1	1.15
Design	3	0.49	–	–
Geotechnical	9	1.48	4	4.60
Bearing	1	0.16	1	1.15
Sturm/Hurrican/Tornado	17	2.79	–	–
Tree Fall	2	0.33	–	–
Miscellaneous/other	6	0.99	1	1.15
Total	609	100	87	100

landslides are related to certain topographies. Seismic loads are related to regions with seismicity.

Table 6.6 and Fig. 6.15 show the causes of bridge collapse whether the bridge overpass a river or a road and railway line. The table indicates that hydraulic caused collapses contribute with nearly 2/3 to all collapses for bridges crossing water. Still collisions (ships) contribute with a value below 10%. In contrast, for bridges overpassing roads and railway lines, hydraulic caused failure contributes with below 10%. One might ask way bridges not crossing water are exposed to hydraulic loads, however many floods have shown that large areas are flooded including roads crossed by bridges. The major contributor to the collapses for the latter group is collision with almost 60%. Interestingly overload is a larger contributor for bridges over water whereas it only contributes minor to bridges over roads and railway lines. Further, fire is a minor contributor for bridges crossing water whereas fire is a larger contributor for bridges crossing roads. Seismic loads reach the same level of contribution for both bridge groups.

The description of the location of the bridge over water can be further refined into bridges over waterways. This may influence the collapse causes. For example in Germany app. 1500 bridges over waterways exist, about 60% of them are exposed to a possible ship impact (Proske 2004).

Table 6.7 shows a list of bridge collapses with fatalities caused by ship impact. However, not all accidents yield to collapses or to fatalities. An extended list can

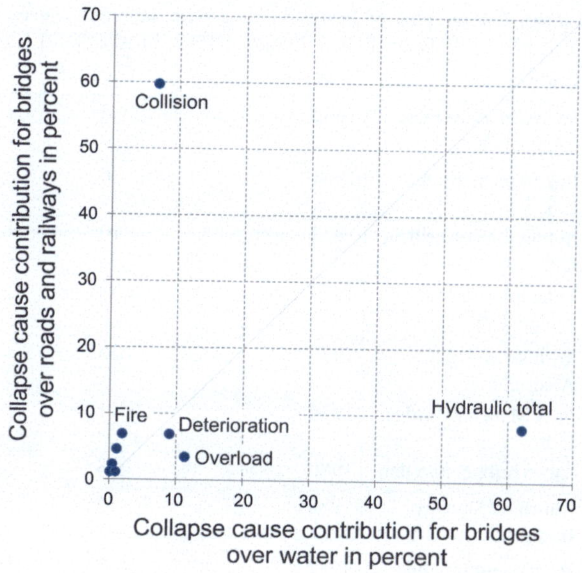

Fig. 6.15 Cause-proportional conditional collapse rate for bridges in the U.S. (Cook 2014)

be found in Frandsen (1983) and Gucma (2015) including collapses without fatalities. Also several ship impacts did not cause the collapse of the bridge (near misses). Examples are the ship impact against the railway bridge Krems in Austria on December 17th 2005 moving the pier app. 2 m (Simandl et al. 2006), against the bridge Segnitz in Germany in 2000 causing a 3 m long crack in the pier (Proske 2009) and the bridge over the Süderelbbrücke (Seipelt et al. 2016). Figures 6.16 and 6.17 show examples of vessel impact against bridges. In some cases, vessel impacts are not an accidental load anymore: the Wuhan Yangtze River Bridge suffered 70 vessel collisions between 1957 and 1999 (Zhang et al. 2016). A systematic study on bridge over the waterway Main in Germany revealed that 55 bridges of the 120 investigated bridges were not able to cover the load by vessel impact (Main-Netz 2009). More details about vessel impact frequencies can be found in Proske (2004) and in Gucma (2015). Examples for the calculation of bridges under vessel impact can be found in Consolazio et al. (2010), Davidson (2010) and Raithel et al. (2011).

Impacts caused by other means of transport are also well known: In June 1998, a railway impact against the bridge in Eschede, Germany, killed 101 people and 88 people were badly injured. An example of the damages of a car impact against a bridge is shown in Fig. 6.18.

Therefore Fig. 6.19 shows the geographic location of the bridges stocks statistically investigated. In general all geographic and climate conditions are covered including Alpine regions and flood prone areas, however bridge collapse data from temperate climate zone is dominating. Currently it is not possible to differentiate the bridge collapse data related to climate conditions such as the Köppen classification system. Climate change may also affect the bridge collapse statistics.

Table 6.7 Examples of bridge failures with fatalities caused by ship impact (extended from Mastaglio 1997; Gucma 2015; Frandsen 1983; Proske 2009)

Bridge name	Year	Fatalities	Ship
Severn river railway bridge, U.K	1960	5	–
Lake Ponchartain, U.S.	1964	6	–
Sidney Lanier bridge, U.S.	1972	10	–
Lake Ponchartain bridge, U.S.	1974	3	–
Tasman bridge, Australia	1975	15	–
Pass Manchac bridge, U.S.	1976	1	–
Tjorn bridge, Sweden	1980	8	M/V Star Clipper
Sunshine Skyway bridge, U.S.	1980	35	M/V Summit Venture
Richemont pipeline bridge, France	1982	7	Set of 2 barges
Sentosa Aerial Tramway, China	1983	7	Dredger (A = 69 m)
Volga river railroad bridge, Russia	1983	176	F/S Aleksandr Suvorov
Claiborn avenue bridge, U.S.	1993	1	M/V Chris
CSX/Amtrak railroad bridge, U.S.	1993	47	n/a
Port Isabel, U.S.	2001	8	Set o barges
Webber-Falls, U.S.	2002	14	Set of 3 barges
Great belt west, Denmark	2005	1	M/V Karen Danilesen
Jiujiang bridge, China	2007	9	n/a
Jintang bridge, China	2008	4	n/a

Although neither the study nor this chapter is directly related to risk assessment, as discussed in Chap. 5, the examples show that bridge collapses occured with up to 200 fatalities per individual collapse (Proske 2004; Proske and van Gelder 2009; Imam and Chryssanthopoulos 2012; Zhang 1993; Biezma and Schanack 2007) and that single natural events may cause the collapse of hundreds of bridges. The latter can cause a strong fluctuation of the collapse frequencies.

Fig. 6.16 Ship impact in Dresden (Germany) at the Albert-Bridge 2015 (*Picture* D. Proske)

6.6.3 Bridge Collapse Fluctuation

Figure 6.20 shows the distribution of the number of bridge collapses in the U.S. between 1989 and 2012 (Wardhana and Hadipriono 2003; Taricska 2014; Hersi 2009) in absolute numbers and Fig. 6.21 in relative numbers related to the maximum value of each data series. The high number of bridge collapses in the year 1989 is related to the Loma Prieta earthquake (Wardhana and Hadipriono 2003). The peak in 1993 is related to a major flood in which the Mississippi and Missouri rivers and their tributaries flooded several Midwest states. Numerous bridge collapses occurred in Iowa, Minnesota, and Missouri (Wardhana and Hadipriono 2003). The peak in 1996 is also related to floods (Wardhana and Hadipriono 2003). The peak in 2005 is related to several floods including Hurricane Katrina (NIST 2006) and the Mid-Atlantic and New England flood in October 2005. Finally the peak in 2011 is mainly related to the Spring Mississippi River Floods (Taricska 2014). It is not known whether the 2011 Mid-Atlantic Flooding contributed. Basically, the pictures confirm the declining trend of the collapse frequency also for the last years. This applies both to the consideration of years with high collapse frequencies and to the exclusion of these years. The years 2009, 2010 and 2012 show exceptional low collapse rates. The pictures do not show any deterioration due to poorer bridge conditions or climate change.

Fig. 6.17 Ship impact against the Mainbridge Lohr 1999 (*Picture* Road Department Würzburg)

Figure 6.22 shows the frequency of the number of bridge collapses per year. Figure 6.22 shows a high fluctuation in the number of bridge failures per year ranging from zero to eleven. This fluctuation is also visible in Figs. 6.20 and 6.21. According to Cook (2014) the standard deviation is relatively high. All figures indicate that single events, which are usually natural disasters with accidental loads, dominate the collapse frequencies. Therefore these figures confirm the conclusions made in the previous section.

Since Fig. 6.20 shows a high uncertainty, the question arises whether the bridge collapse frequency is still a mild distribution such as the normal distribution or the family of extreme value distributions or already a wild distribution. Mild distributions can be sufficiently described by mean value and standard deviation; in some cases higher order moments or upper or lower bounds are required. In contrast wild or Lévy-distributions show an infinite or non-converging mean value. Distribution parameters for such distributions are difficult to obtain.

Figure 6.23 shows a Kesten process as an example of such a wild distribution (Sornette 2000). Whereas the empirical mean value would be below 100, extreme values up to nearly two thousand can be seen in this example. The question rises whether we find such extreme values in the real world.

If we consider only the tsunamis 2004 in Southeast Asia and 2011 in Japan, we indeed find such huge numbers of damaged and collapsed bridges. Cluff (2007) notes that hundreds of bridges were swept away during the 2004 tsunami in Aceh (Indonesia) alone and Unjoh (2005) notes that more than hundred bridges were swept away in the same region. Shoji and Moriyama (2007) describe the tsunami

Fig. 6.18 Arch bridge after a car impact (*Picture* D. Proske)

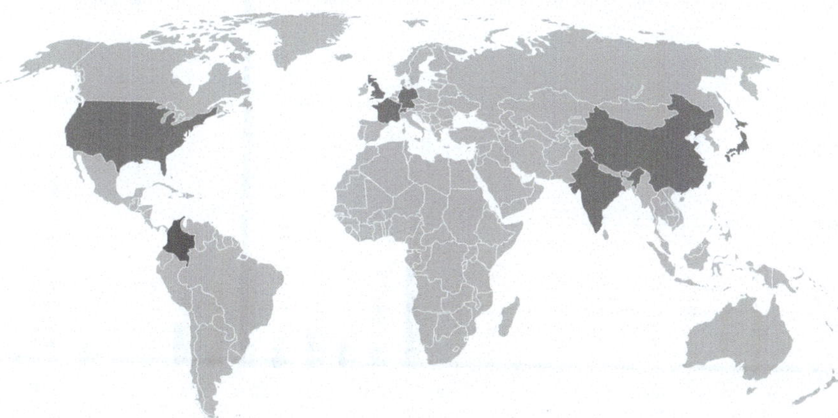

Fig. 6.19 Geographic location of the regions in the studies used

fragility estimation for bridges in Sri Lanka based on the tsunami event 2004. As mentioned before, the earthquake and tsunami 2011 in Japan destroyed more than 300 bridges (Maryama et al. 2012; Akiyama et al. 2012). Kosa (2014) describes in detail the collapse and damages of bridges during the tsunami 2011 in Japan. Yim

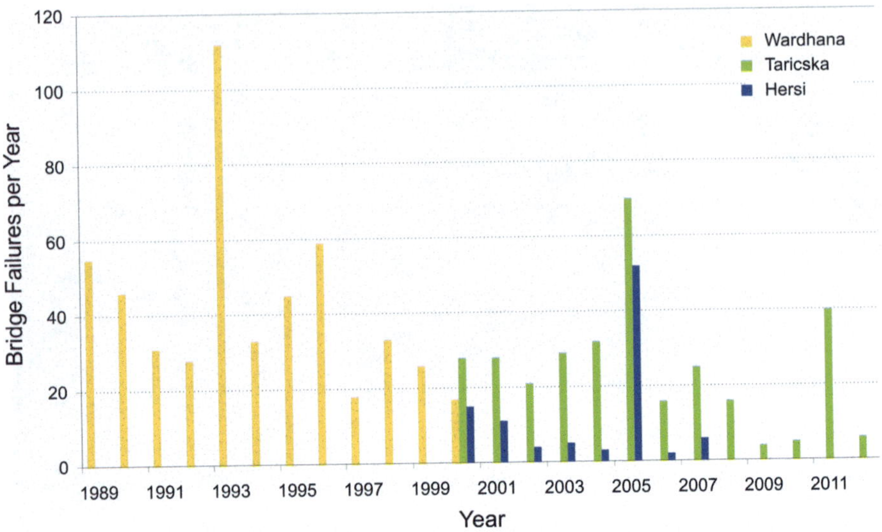

Fig. 6.20 Absolute bridge failures per year according to Wardhana and Hadipriono (2003), Taricska (2014), Hersi (2009)

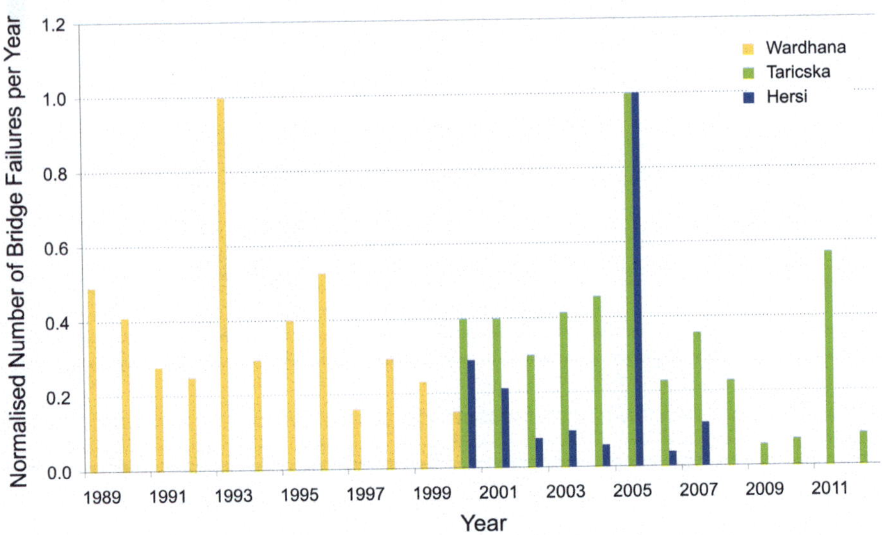

Fig. 6.21 Normalised bridge failures per year according to Wardhana and Hadipriono (2003), Taricska (2014), Hersi (2009)

and Azadbakht (2013) explain the estimation of tsunami impact forces for bridges in California.

We conclude that the mean value given for the collapse frequency of bridges has a limited meaning since large single events may dominate the results. In years with

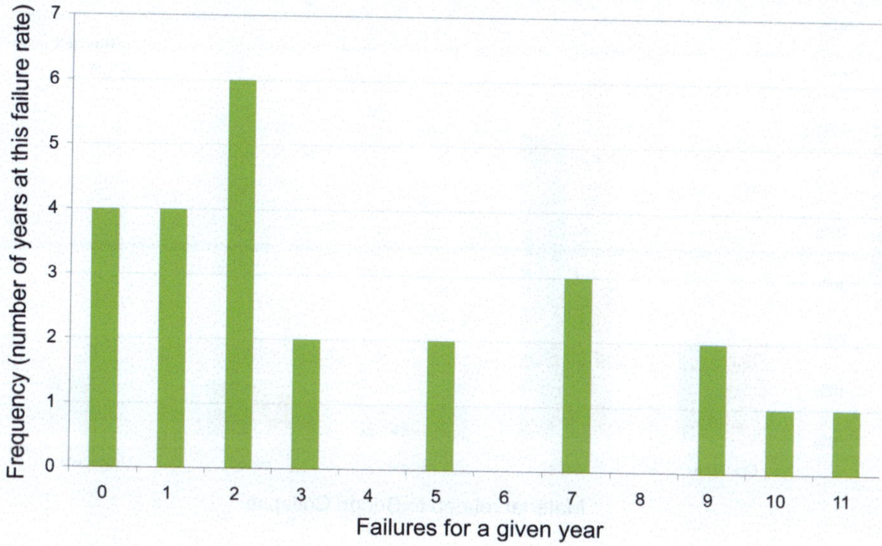

Fig. 6.22 Bridge failure frequencies in the years from 1987 to 2011 (Cook 2014)

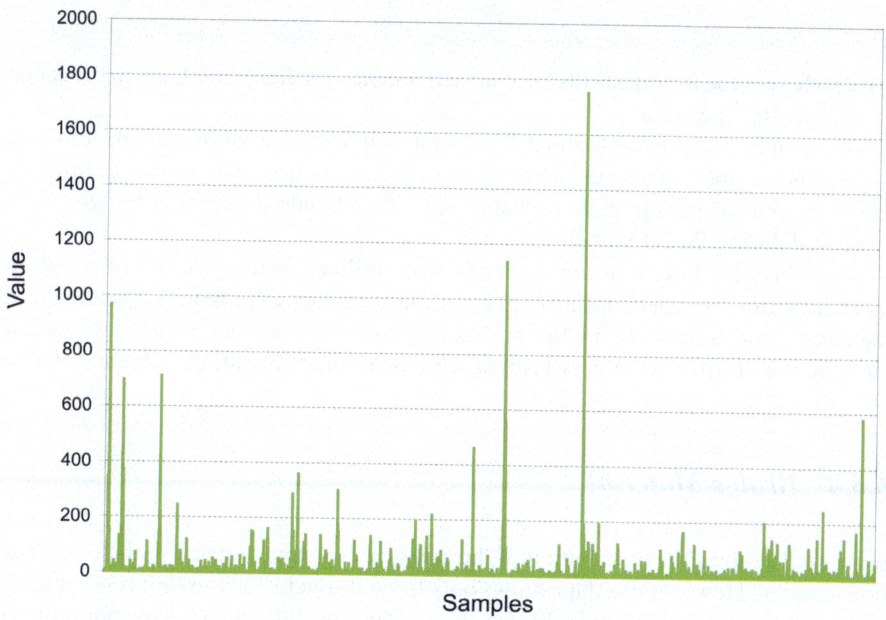

Fig. 6.23 Example of a Kesten-process with very extreme values

extreme events the number of bridge collapses may be several orders of magnitude larger than in years without extreme events. The collapse frequency for bridges exposed to the tsunami wave in 2011 in Japan reached up to 10% (Maruyama et al.

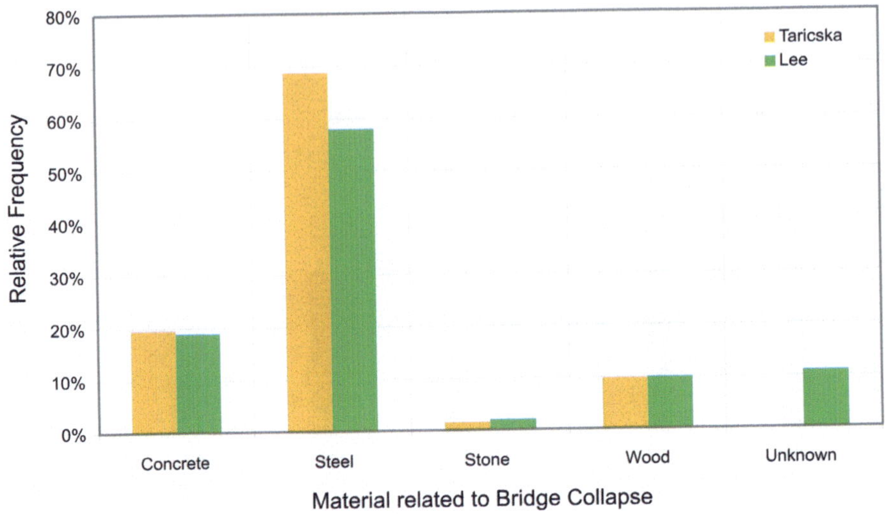

Fig. 6.24 Relative frequency of bridge collapse related to the construction material based on Taricska (2014); Lee et al. (2013a)

2012) whereas the average bridge collapse frequency for Japan might be in the range of 10^{-5} to 10^{-4} per year.

At the end of World War II about 20% of the buildings were destroyed in Germany. In absolute numbers this would correspond to the destruction of thousands of bridges. Looking on a worldwide scale probably more than hundred thousand bridges were destroyed during World War II.

Nowadays for bridge designers loads from military action are out of scope in most countries. However during history military actions were indeed considered in the design, for example by including blasting chambers in piers, including wooden parts for fire destruction or drawbridges, casemates in abutments etc. (Mende 2016).

6.6.4 *Bridge Material*

Figure 6.24 shows the distribution of the collapse frequency related to the construction material. However this data allows only limited conclusions since it is not related to the material ratio of the entire bridge stock. Therefore the data has to be normalized to the bridge stock ratios of the materials. This is carried out for the U.S. data and shown in Fig. 6.25.

Figure 6.25 allows several conclusions: The collapse frequency of steel bridges is significantly greater than for concrete bridges (6:1). This conclusion and the value is not only based on Fig. 6.24 according to Cook (2014), but also strongly supported by Lee et al. (2013a). If the steel bridge collapses are more frequent than concrete bridge

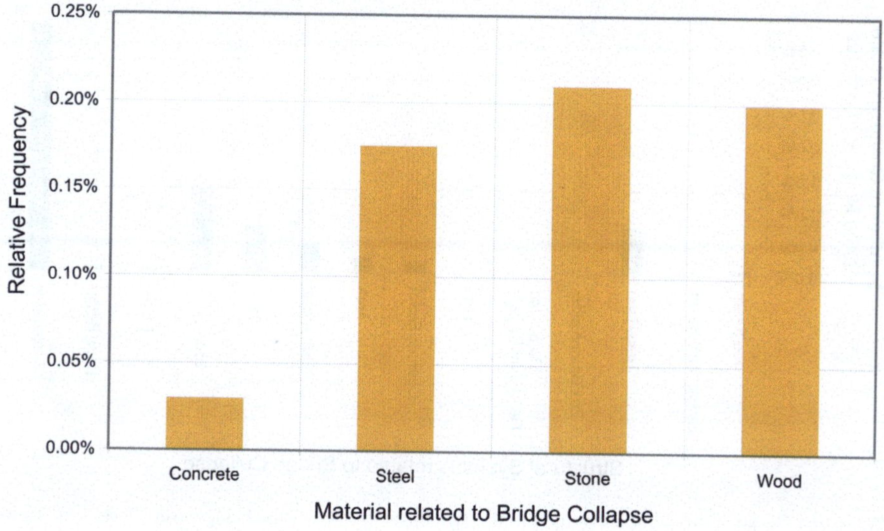

Fig. 6.25 Relative frequency of bridge collapse related to the construction material (normalised) based on Taricska (2014)

collapses, the collapse frequency can be lower in countries with a higher concrete bridge stock. This fact is indeed true for Germany where the ratio of reinforced and pre-stressed concrete bridges to the overall bridge stock seems to be higher then in the U.S. (see Chap. 2). Based on the data available the collapse frequency in Germany should be half the value in the U.S. However, this is a very simplistic approach since many possible factors such as

- different safety requirements,
- different extreme weather conditions (e.g. regular hurricanes in the U.S.),
- different major means of transport,
- different preferred construction types and technologies,
- different maintenance strategies and
- different age distribution

may affect the comparison of the collapse frequency between two regions or time periods.

Furthermore, the contribution of masonry is very low in the non-normalised figure and surprisingly high in the normalized figure. The first observation suggests that masonry bridges contribute only minor to the bridge stock. Since Proske and van Gelder (2009) have shown that masonry and natural stone arch bridges are usually very robust for all types of vertical live loads, the second observation means bridge failure will probably be caused by flooding and scour (see for example Figs. 6.9 and 6.14).

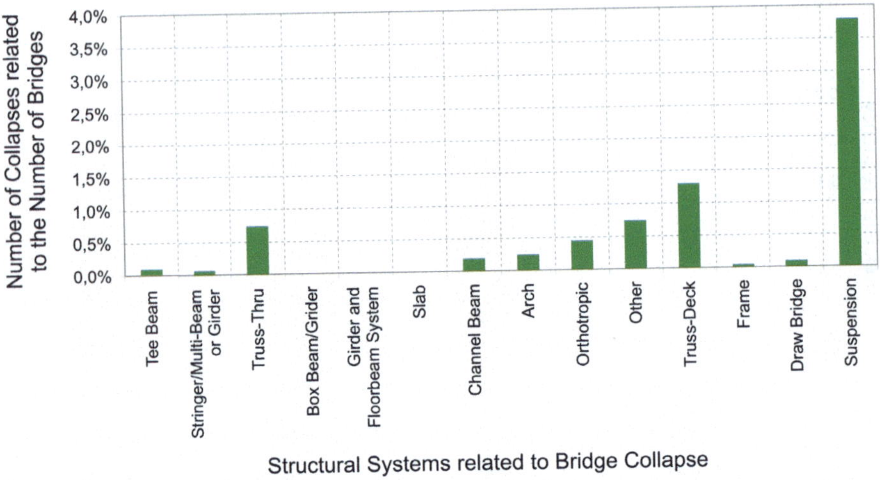

Fig. 6.26 Collapse frequencies related to the structural system of the bridge considering the ratio of the structural systems on the overall bridge stock (normalised) according to Taricska (2014)

6.6.5 Bridge Structural System

Besides the construction material, Fig. 6.26 shows the normalised contribution of structural systems and bridge types to the collapse frequencies. Since the data is normalized based on the contribution of the bridge type to the overall bridge stock, the figure directly shows the weakness of some bridge types. First of all, suspension bridges seem to be extremely unreliable (see Sect. 6.9). This would be in compliance with the Fig. 6.2 indicating high collapse frequencies of suspension bridges in the U.S. in the beginning of the 20th century. The second largest contribution comes from Truss bridges in both types, Truss-Thru and Truss-Deck. Putting Truss-Deck and Truss-Thru bridges together, Truss bridges reach more than 2.5%.

Masonry and concrete arch bridge failure is usually related to accidental loads such as floods as mentioned before (Proske and van Gelder 2009). Steel arch bridge failure is mainly related to fatigue (Imam and Chryssanthopoulos 2012).

Non-normalised collapse data related to structural systems is given in Sharma and Mohan (2011), Wardhana and Hadipriono (2003) and Taricska (2014). Imam and Chryssanthopoulos (2012) note that 35% of all metallic bridge collapses are related to Truss bridges, however they only contribute with 29% to the structural configuration. This confirms the conclusions from Fig. 6.26. All other structural configurations show a lower contribution to the collapse numbers compared to the bridge stock structural configuration. The best ratio of collapse contributions to bridge stock ratio for steel bridges show girder bridges (Imam and Chryssanthopoulos 2012).

Deng et al. (2016) have related the causes of collapses to the type of structural system and construction material. They have shown that not every bridge type is vulnerable to every cause discussed in this chapter so far. The results are summarized in

Table 6.8 Most common cause of collapse of different structural types of bridges (Deng et al. 2016)

Type of bridge	Most vulnerable cause
Beam	Flood, scour, earthquake, collision, overloading
Masonry arch	Flood, scour, overloading, earthquake
Steel arch	Overloading, wind
Steel truss	Overloading, fatigue
Flexible long-span	Wind

Table 6.8. However the author can not confirm all conclusions, for example masonry arch bridges are usually not vulnerable to overloading since they show an excellent load bearing behaviour under vertical loading (Proske and van Gelder 2009): most masonry arch bridges are demolished due to limited serviceability and collapsed due to flooding and scour. There are also some studies regarding the behaviour of arch bridges under seismic loading which do not in general indicate a significant weakness of masonry arch bridges under seismic loading.

6.6.6 Bridge Age Distribution

One can argue that all the potential causes are strongly related to the age of the bridges. Figure 6.27 shows collapse frequency data based on Cook (2014), Taricska (2014) and Lee et al. (2013a, b). This diagram does not confirm this assumption even there is a peak in Cooks (2014) data at the age of 30 years. This peak is neither confirmed by the other authors nor by the Figs. 6.2, 6.3, 6.4, 6.5 and 6.6. Additionally Fig. 6.28, which shows the time dependency of different causes of collapses, indicates that limited knowledge is constantly decreasing and design errors are constant over the last 50 years. Therefore the peak is not related to design changes in the last decades, but the peak may be related to a single accidental event which caused the reconstruction of a large group of bridges several 30 years ago and the bridge were exposed again 30 years later (see also Sect. 6.9 and Fig. 7.1).

Figure 6.29 shows a Boxplot of the bridge ages related to some causes of collapse. The diagram is a strong indication that overload is more related to older bridges whereas bridges of all ages are affected by hydraulic events and collisions are more related to younger bridges. This does not mean that young bridges are weaker then the old ones, it just means that newly built bridges over rivers are sooner exposed to a collapse-causing flood then to a collapse-causing overload. Since the flood loads show great uncertainty and the traffic load is usually growing this fact is not surprising. Overall the median age of the bridge population is slightly younger then the age of collapsed bridges with 55 years (Montalvo and Cook 2017). Hence weak aging effects appear to be visible.

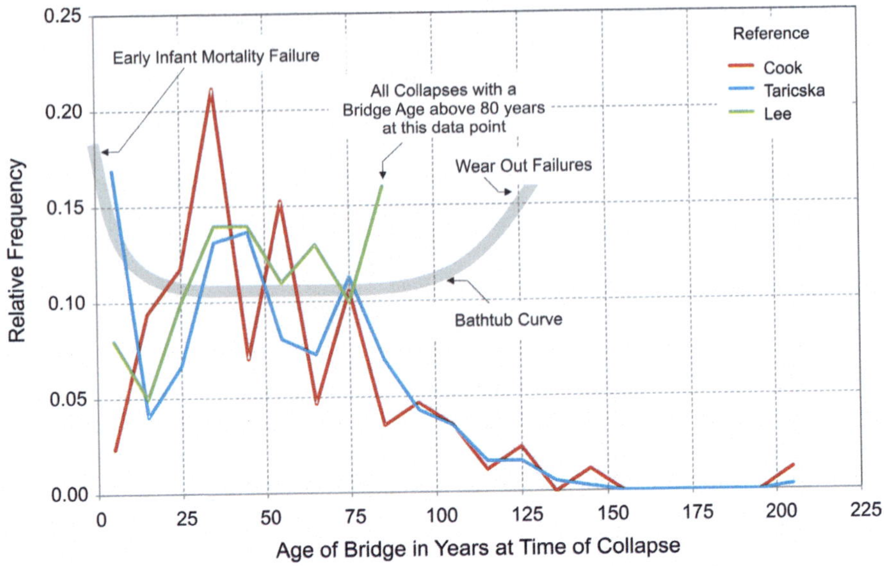

Fig. 6.27 Relative frequency of bridge collapse related to the bridge age (based on Taricska 2014; Lee et al. 2013a; Cook 2014)

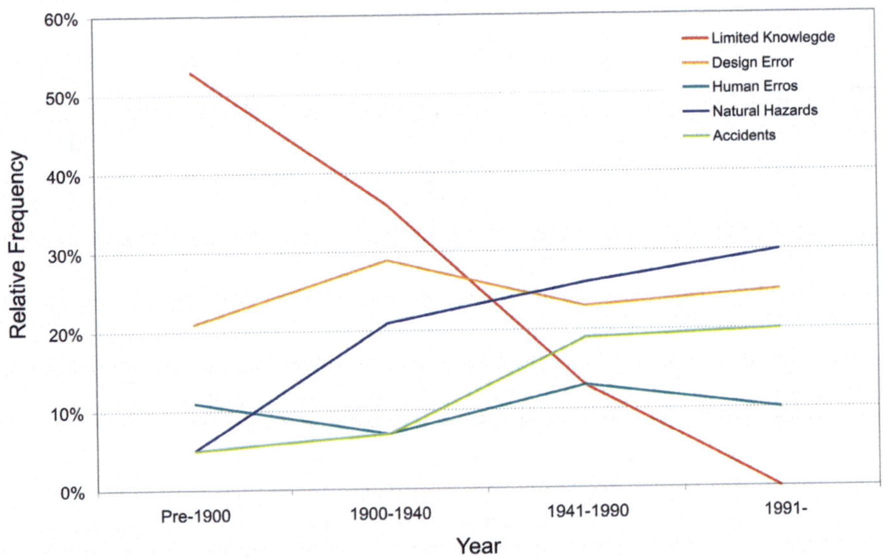

Fig. 6.28 Time dependency of causes of bridge collapses for steel bridges according to Imam and Chryssanthopoulos (2012)

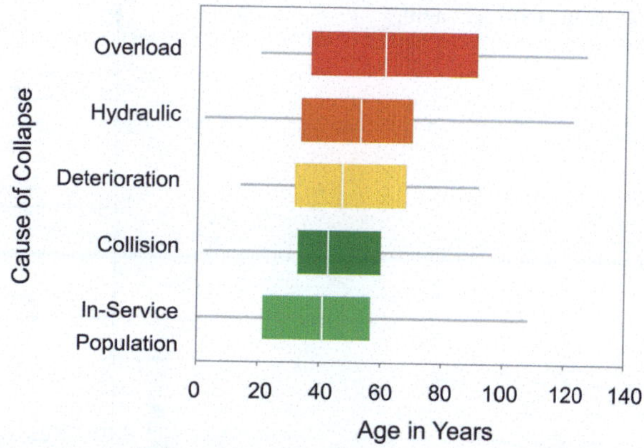

Fig. 6.29 Box-plot of bridge age and cause of collapse (Montalvo and Cook 2017)

6.7 Prediction of Future Collapse Frequencies

Based on the conclusions so far we should be able to predict future bridge collapse frequencies. The simplest approach would be the extension of the decreasing trend into the future. Since the number of refurbished bridges is rather low in industrialised countries, usually a low one digit percent value, the average age of the bridge stock will remain or slightly increase. In the last years in Germany the average age of the capital stock of infrastructure has increased (Boysen-Hogrefe et al. 2013).

Therefore the effects of aging and the countermeasure—maintenance—become more and more dominating. Figure 6.30 shows various theoretical models of aging of technical products, to which bridges belong. The bathtub curve has already been used in Fig. 6.27; for further information regarding the bathtub curve see Eberlin and Hock (2014). Based on this consideration the overall collapse frequency will increase at a certain point in time, except we consider no aging concepts. However, observations on the bridges indicate an average worsening of the bridge conditions (see Figs. 6.31, 6.32, 6.33 and 6.34). Therefore the concept of no aging can be excluded if there exists a relationship between the bridge condition and the collapse frequency and the probability of failure respectively.

First, using Fig. 6.33 a relationship between the age distribution and the bridge condition can be developed. The relationship is assumed as follows:

$$C(t) = 1.8 + 0.0081 \cdot t$$

with t as age in years, for example 80 year etc.

Figure 6.34 shows a comparable function for U.S. non-highway bridges.

Based on the figures and the equation we can expect a further decrease of the bridge conditions in Germany.

Based on the idea of Davis-Mcdaniel (2011) one can relate the bridge condition to a probability of failure (Fig. 6.35). Although the application of the numbers from

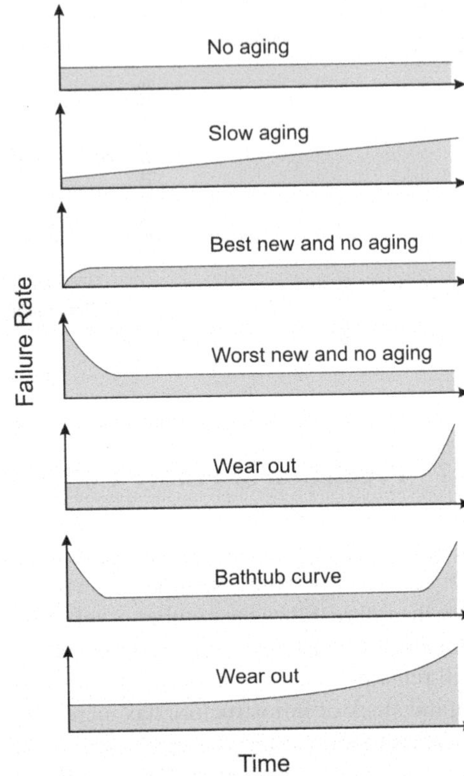

Fig. 6.30 Different aging theories

Davis-Mcdaniel (2011) is not exactly what was done in the publication, it may be used as a preliminary approach. Results are shown in Tables 6.9 and 6.10. In general, the overall number given in Table 6.9 is far too high (more than fifty bridge collapses per year for the German highway bridges). The result is a strong indicator that either the used probability of failure is conservative, that the maintenance has a very significant effect on the final collapse frequency and that our time period for the estimation of the collapse frequency is simply too short. If the probabilities of failure are decreased by one order of magnitude as shown in Table 6.10, the number of bridge collapses decreases to five only, a much more realistic number.

Figures 6.30, 6.31, 6.32 and 6.33 indicate that the conditions of bridges in industrialized countries decrease. However, it is unclear whether this yields to an increased collapse frequency. Furthermore if the decrease of the bridge conditions in the last decade would increase the probability of failure for dead and live load by only about 10%; it would not be visible in the data due to the large contribution of accidental loads to the collapses which may be affected only minor to declined conditions of the bridges.

Another approach for the prediction of future developments is the consideration of the target or acceptance values for existing bridges. The target probability of failure

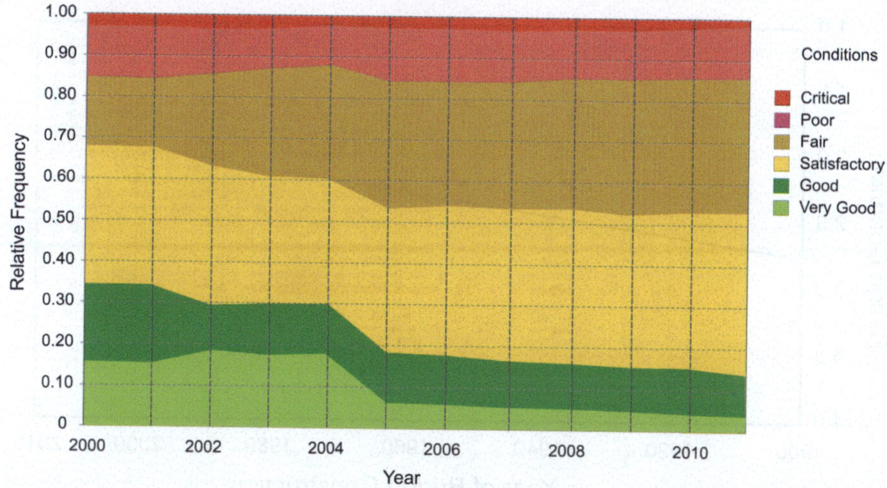

Fig. 6.31 General trend of the bridge conditions on highways and federal roads in Germany (BMVBS 2013, BAST 2016; BAST 2017)

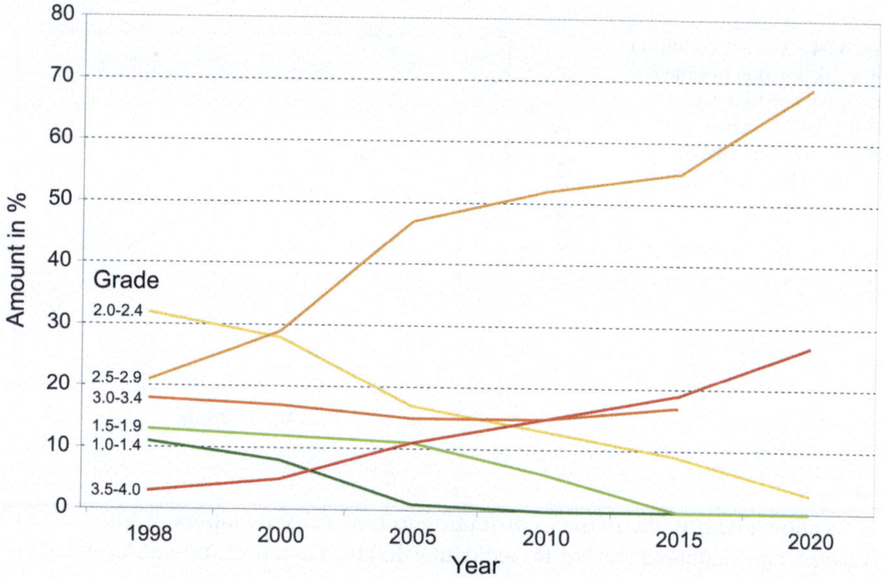

Fig. 6.32 Expected condition development of German highway bridges without enhanced maintenance (1.0 very good, 4.0 insufficient) based on Budelmann et al. (2013)

increases constantly with age for existing bridges (see Chap. 5). One can compare this time-dependent target probability of failure with the observed collapse frequency as shown in Fig. 6.36. Based on this consideration the observed collapse frequency would increase since the replacement rate is too low.

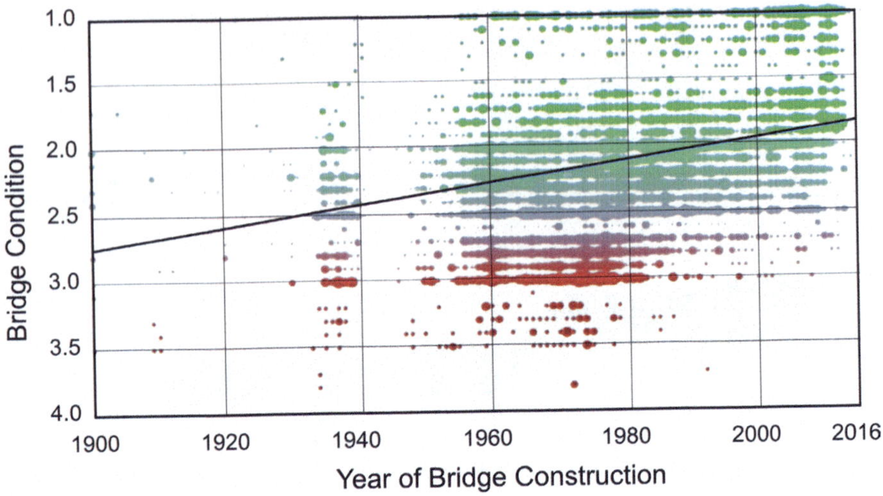

Fig. 6.33 Condition of bridges as function of the year of construction (1.0 very good, 4.0 insufficient) based on Nagel et al. (2016)

Fig. 6.34 Average condition rating as function of time for non-interstate highway bridges (9.0 excellent, 3.0 serious, 0 failure) based on Bolukbasi et al. (2004)

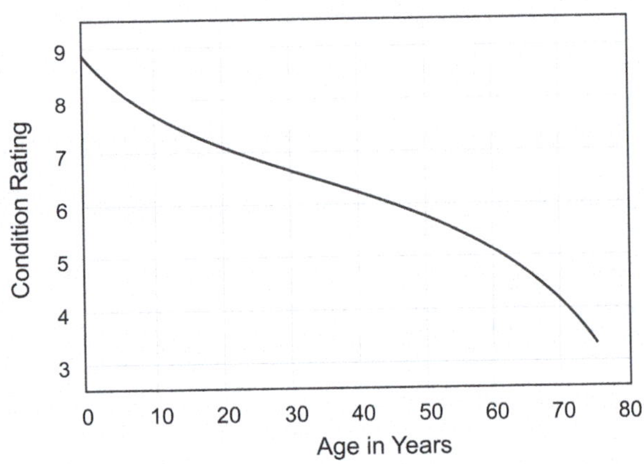

As shown before, the major contribution to the bridge collapses is not related to overload and maintenance but to accidental loads. Therefore mitigation measures related to the accidental loads, such as new flood plains in a catchment area and new dam management regulations respectively may have a stronger effect on the bridge collapse frequencies as maintenance optimisation and aging effects. This may also be true for other major accidental causes such as fender systems as ship impact protection systems, better qualification and training of drivers, automatization, increased police control frequencies or simply the development of the traffic volume.

The large contribution from accidental loads may yield to the fact that developments in the related fields are earlier and easier to identify in the bridge collapse data

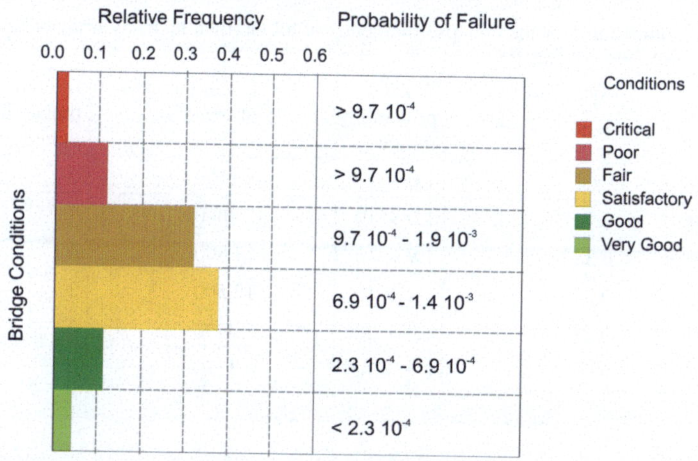

Fig. 6.35 Relationship between bridge conditions and probability of failure

Table 6.9 Computation of the collapse frequency of all German highway bridges based on the concept explained above (Part I)

Age	Age distribution	Related probability of failure	Overall number of bridges	Overall probability of failure
1900–1940	0.04	1.0×10^{-3}	1560	2
1940–1960	0.12	1.0×10^{-3}	4680	5
1960–1980	0.38	1.9×10^{-3}	14,800	28
1980–2000	0.32	1.4×10^{-3}	12,500	18
2000–2010	0.11	6.9×10^{-4}	4290	3
>2010	0.03	2.3×10^{-4}	1170	0
Sum				55

compared to maintenance and aging changes. For example climate change affecting the return period of extreme floods and storm surges may become visible in the collapse data long before trends of maintenance are visible.

It can be summarized that several factors may influence the future development of bridge collapse frequencies such as:

- change in the safety requirements, for example as consequence of an accident,
- change of natural hazards, e.g. climate change, flood prevention measures,
- changes in the means of transport, longer trucks, heavier trucks (for example mass application of electric vehicles).
- changes in the construction material (for example application of textile or carbon reinforcement), new construction technologies and
- changes in the maintenance strategies.

Table 6.10 Computation of the collapse frequency of all German highway bridges based on the concept explained above (Part II)

Age	Age distribution	Related probability of failure	Overall number of bridges	Overall probability of failure
1900–1940	0.04	1.0×10^{-4}	1560	0
1940–1960	0.12	1.0×10^{-4}	4680	0
1960–1980	0.38	1.9×10^{-4}	14,800	3
1980–2000	0.32	1.4×10^{-4}	12,500	2
2000–2010	0.11	6.9×10^{-5}	4290	0
>2010	0.03	2.3×10^{-5}	1170	0
Sum				5

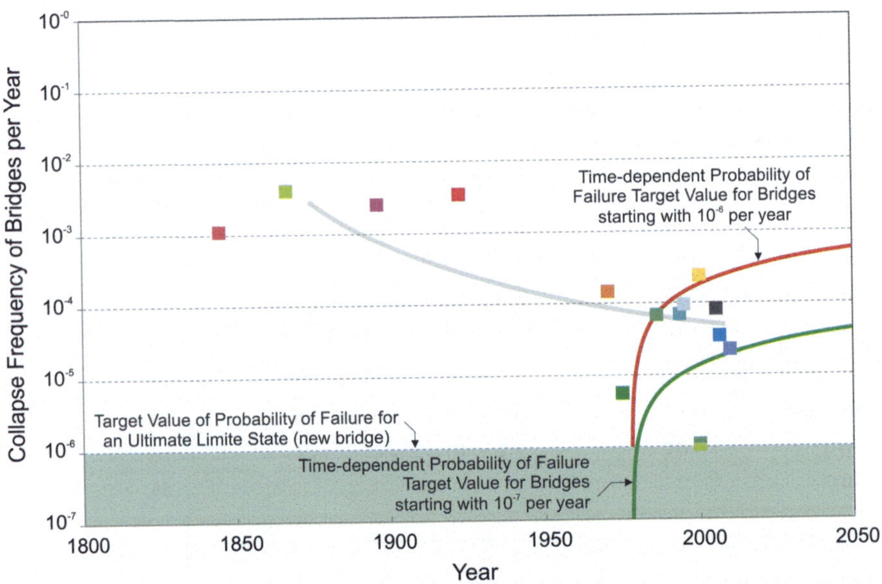

Fig. 6.36 Comparison of the decreasing trend of the observed collapse frequency and some acceptable probability of failure values for a new bridge built in the 1970s

6.8 Comparison of Target Values and Failure Probabilities

In Chap. 5 target values for acceptable probabilities of failure either for individual limit states or for entire structures were given. Usually the values were in the range of 10^{-6} per year for the ultimate limit states. However, this number is related to new bridges. Considering the target values for existing structures, with are significantly higher compared to targets for new structures, they confirm the collapse frequency values. For a thirty years old bridge the target probability of failure is between

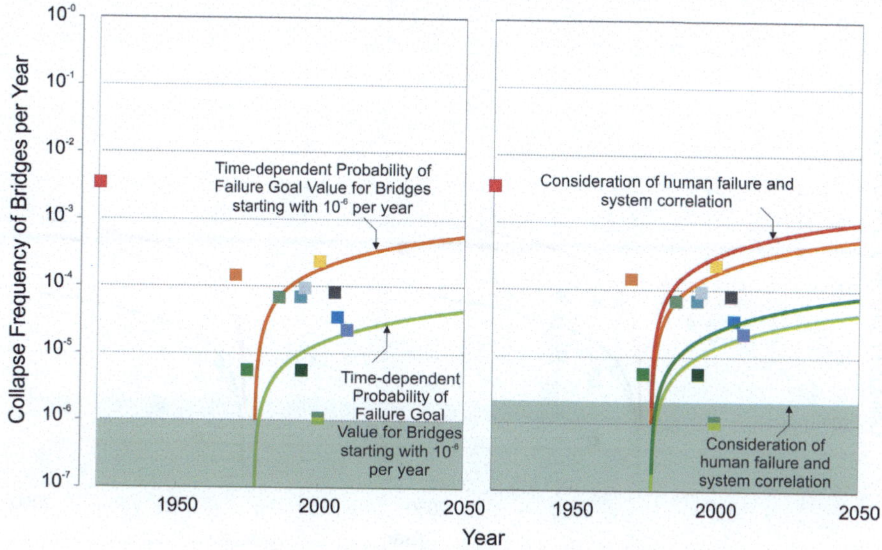

Fig. 6.37 Relationship between target probabilities of failure and collapse frequencies (case I)

3.2×10^{-5} and 2.3×10^{-4} per year. In Fig. 6.37 (left) the time-dependent target values according to SIA 269 (2007) are shown. Most of the observed collapse frequencies are already covered by the SIA 269 curve. These numbers are probably very conservative since presumably the probability of failure does not increase as fast as it is assumed in the SIA 269 curve.

If additional effects such as human failure or system correlation are considered, the time-dependent target value for the bridges increases in average by 30% to 1.3×10^{-6} per year as target. As Fig. 6.37 (right) shows, all data points related to the period after 1970 are covered.

As seen in Fig. 6.6, the collapse frequency of bridges is dominated by accidental loads, mainly related to floods. The target values given usually refer to combinations of dead and live loads. Using the data from Fig. 6.6 one can estimate the collapse frequency related to overload, here used as a substitute for the dead and live load conditions. This is shown in Fig. 6.38 (left). As a consequence, the SIA 269 curve covers all collapse data points with large margins (Fig. 6.38 right).

It can be concluded that the effective collapse frequency of bridges for the dead and live load conditions expressed as overload is probably in the range of 10^{-6} to 10^{-5} per year. Due to the low contribution of overload to the overall collapse frequency and due to the large effects of single events such as floods, changes of the collapse frequency of bridges due to changing dead and live loads are difficult to identify.

However, for some accidental loads the target values are higher, in the range of 10^{-2} per loading. Since the collapse frequencies are governed by accidental loads, larger collapse frequencies than target probabilities of failure for live and dead loads

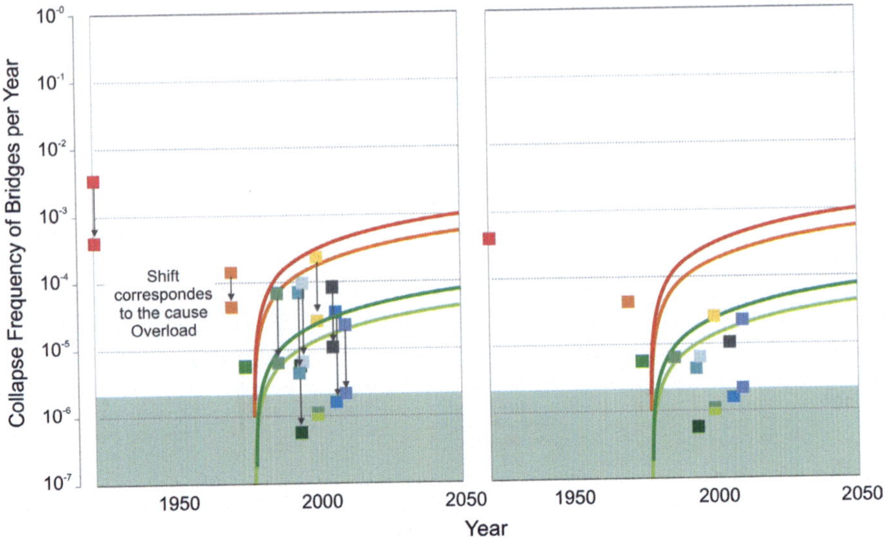

Fig. 6.38 Relationship between target probabilities of failure and collapse frequencies (case II)

should be expected for such rare accidental loads. However there return period is very large and limits the effects on the collapse frequency.

Finally Fig. 6.39 indicates that maintenance may significantly affect the target probabilities of failure. The selection of the starting point of the renewed probability of failure function using app. 50 years is based on recommendations in Austria (ASFINAG) regulating the return period of major maintenance works (40 years). Further details are given in Petschacher (2007), FSV (2017) and Binder and Strauss (2017). However even then, the observed collapse frequencies are covered.

As already shown in Fig. 6.40, several different factors influence the observed collapse frequencies and the target probabilities of failure, if the factors are also included in the latter. This conclusion is in compliance with the original statement, that the collapse frequencies and the probabilities of failure must not be compared. However, the time-dependent target probabilities must cover the diversity of time-dependent probabilities of failure as shown for some examples in Fig. 6.41 otherwise the target values are useless. Indeed a diversity of target probabilities indicates this diversity of time-dependent probabilities of failure as shown in Fig. 6.42. Especially the end of life target probabilities show extremely high values (considering the fact that the lifetime of bridges is often extended due to financial limitations) covering all possible developments of probabilities of failure over time.

In Fig. 6.43 (right) the collapse frequency data points (coloured squares) are compared with several probabilities of failure values for existing individual bridges (red dots). These probabilities of failure try to model each individual bridge as realistically as possible. The selection of the individual bridges was done with the aim of representative samples. Figure 6.43 shows a good consistency of the observed aver-

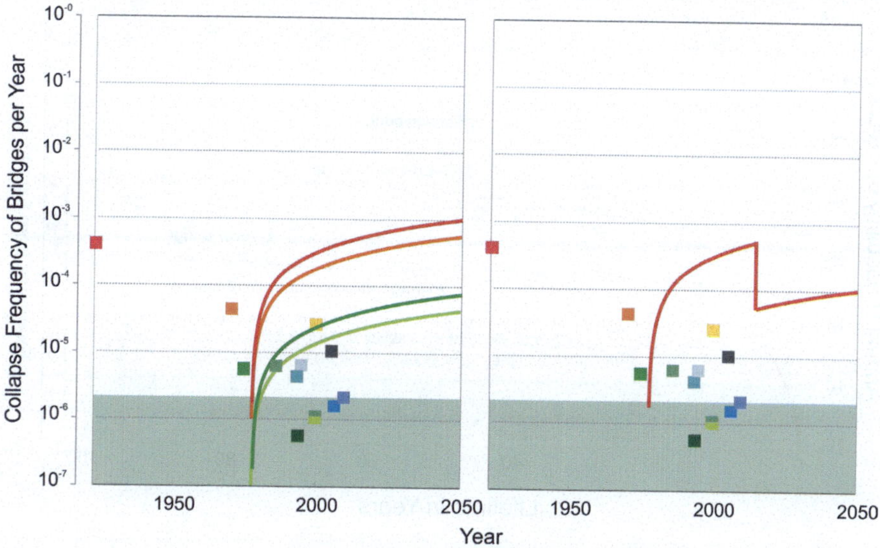

Fig. 6.39 Relationship between target probabilities of failure and collapse frequencies (case III)

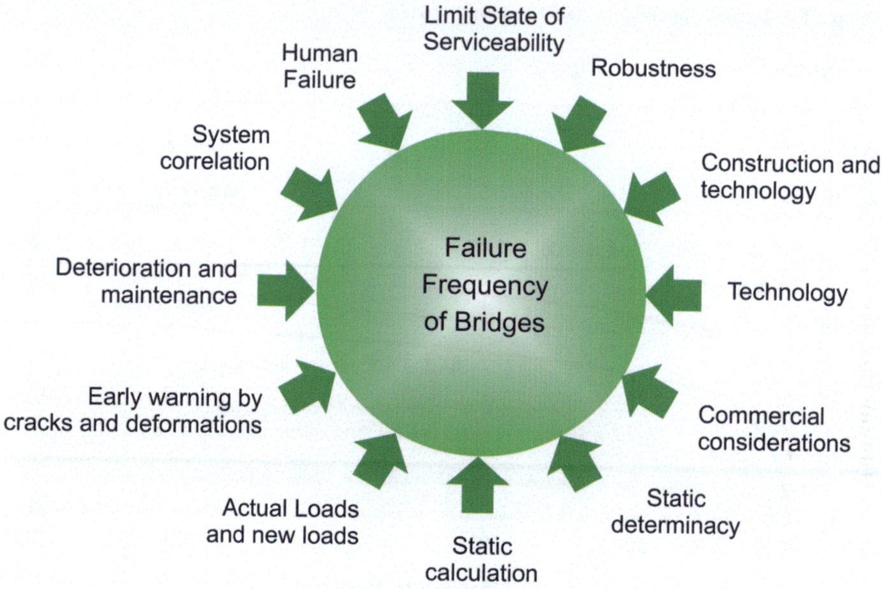

Fig. 6.40 Potential factors influencing the observed failure frequency of bridges

age frequency numbers and the computed individudal probability numbers; perhaps slightly smaller observed values. The observed collapse frequencies and the computed probabilities of failure are also compared in Fig. 6.44 in terms of a histogram.

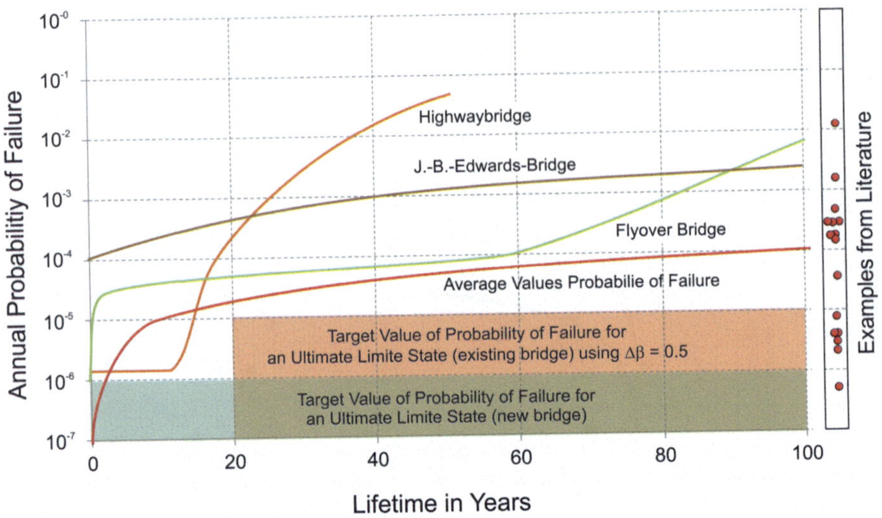

Fig. 6.41 Development of the time-dependent probabilities of failure (represented by lines) for three examples according to Davis-Mcdaniel (2011), Frangopol and Okasha (2008), Schneider et al. (2015) and several time-independent probabilities of failure (represented by red dots), see for example Proske and van Gelder (2009) and Ang (2013)

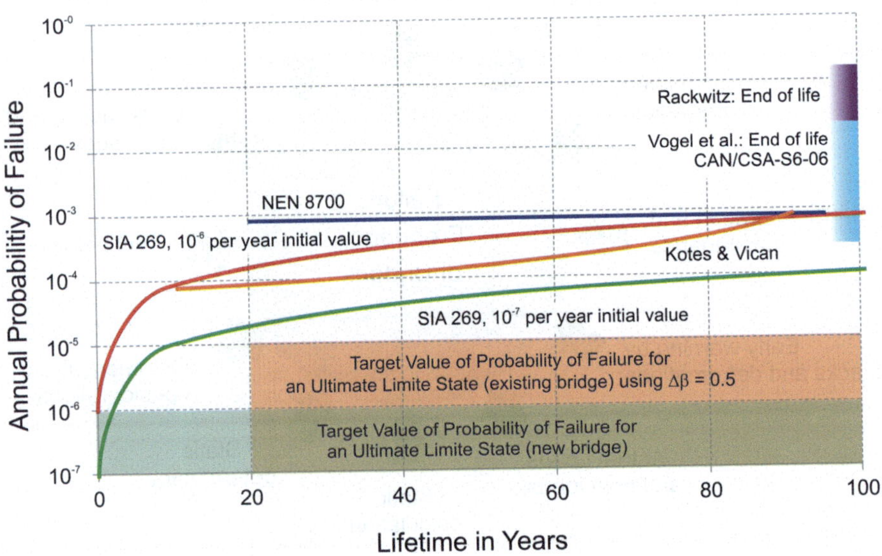

Fig. 6.42 Development of the time-dependent target probabilities of failure, mainly taken from Vogel et al. (2009), Fischer (2010), SIA 269 (2007) and Kotes and Vican (2012)

Although the histogram is not a perfect tool for comparison due to the selection of classes, it is a strong indicator that no systematic difference between the collapse

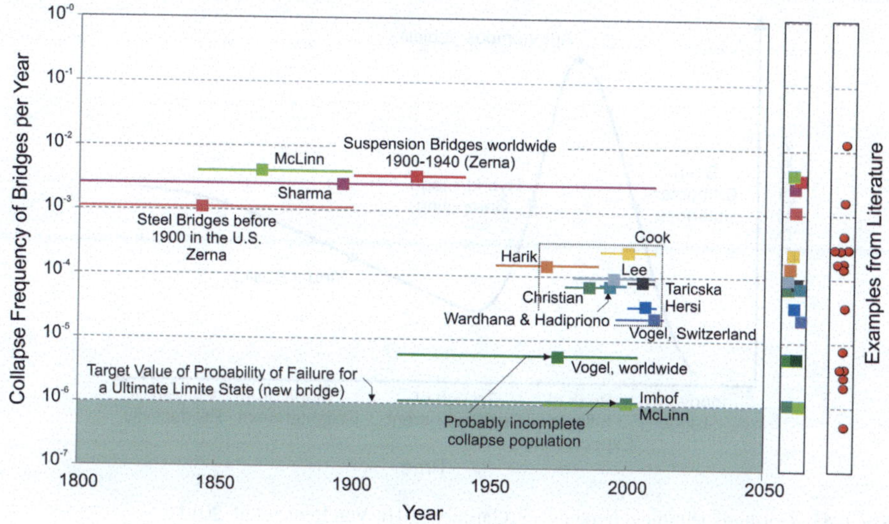

Fig. 6.43 Comparison of the observed collapse frequencies and computed probabilities of failure

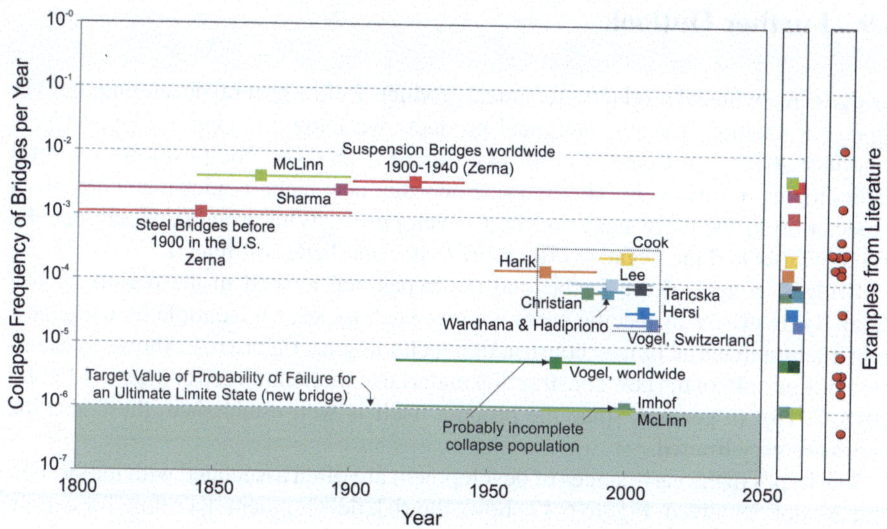

Fig. 6.44 Histogram of the observed collapse frequencies and computed probabilities of failure

frequencies and the probabilities of failure can be observed. The probabilities of failure show a mean value of 9×10^{-4} and a median of 1.5×10^{-4} per year whereas the collapse frequencies show a mean value of 6×10^{-4} and a median of 8.5×10^{-5} per year. The standard deviation of the probabilities of failure is slightly higher then the value for the collapse frequencies but in general, the statistical parameters do not significantly differ.

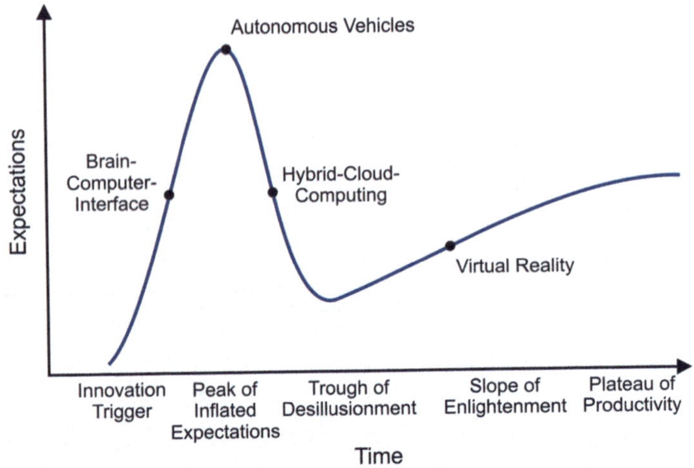

Fig. 6.45 Gardners technical hype cycle (Gardner 2016; Van Lente et al. 2011)

6.9 Further Outlook

Bridges are technical products. Technical products follow general developments over time. For example for new technical products we know the Gardner Hype Cycle (Gardner 2016; Van Lente et al. 2011). Figure 6.45 shows the position of several technologies in this cycle. The figure shows that new technologies reach a peak of expectation in the early stages of their development. With growing experience the true potential and the negative effects are better and better estimated.

Bridges in general are far beyond the hypes and located in the region of productivity. However in specific cases we face such hypes, for example by using new construction material or new construction technologies. Figure 6.46 shows as example a bridge built of the new construction material textile reinforced concrete (Michler 2013, 2016). In general, since the product itself is well established, these specific hypes are very limited.

The hypes in the early stages of development are often associated with major risks that are not perceived. Figure 6.47 shows the risk development over time for certain products. It shows that in the beginning the objective risk increases (position of data point medicine residue in ground water in the diagram in Fig. 6.47). This fact only takes into account the risks related to the product itself, many products however yield to an overall decrease of risks by substituting products with higher risks. At a certain point or usually after an accident or catastrophe related to the product, the subjective risk awareness increases and safety measures are installed (position of the data point mobile phone in the diagram in Fig. 6.47). This yields to a decrease of the objective risk and the subjective risk decreases delayed. As shown in this document, the objective risk of structures in general and for bridges specifically is very low (position of the data point chlorofluorocarbon in the diagram in Fig. 6.47).

Fig. 6.46 Textile reinforced concrete bridge in Germany (*Picture* H. Michler)

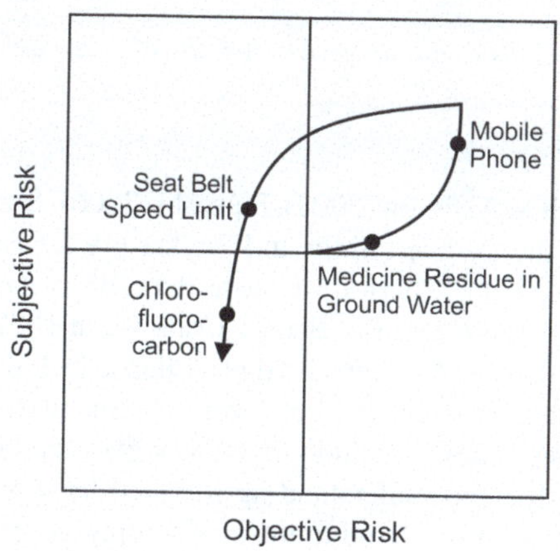

Fig. 6.47 Risk cycle adapted from Metzner (2002)

There are also no indications related to higher subjective risk awareness. Structures are usually considered as safe by the public.

This fact is confirmed by the general usage of the technology. Figure 6.48 shows the development of the commercial aviation subject to the risk and the part of the population using this technology. Whereas the application of some technologies is under discussion, such as Nuclear Power or genetically modified food, the use of bridges never seems to be a point of discussion. In fact, people complain if bridge access is limited due to safety concerns.

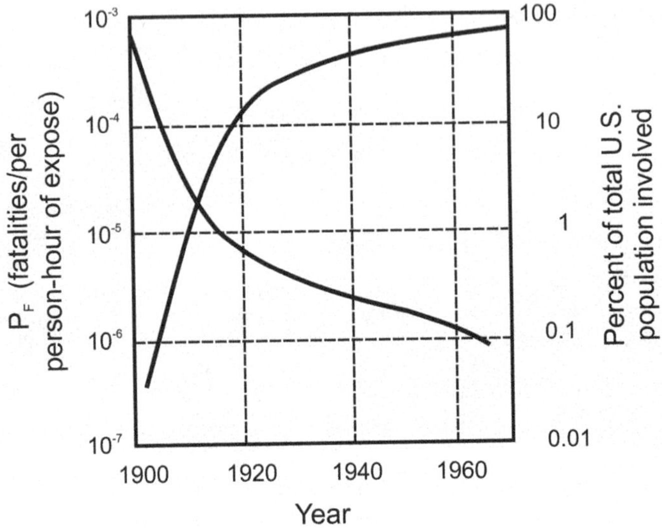

Fig. 6.48 Development of FAR over time and development of acceptance of air traffic technology (Starr 1969)

Several authors such as Akesson (2008), Petroski (2006), Brady (2013) and Steedman (2010) refer to a theory from Sibly and Walker (1977) considering major collapse events of bridges over a period of more than 100 years. They consider the Dee Bridge collapse in 1847, the Tay Bridge Collapse in 1879, the Quebec Bridge Collapse in 1907, the Tacoma Narrows Bridge Collapse in 1940 and the Box-Girder Bridge Failures between 1969 and 1971 as turning points in design and construction development of bridges. They conclude that every 30 years a paradigm shift with respect to the knowledge and understanding initiated by a bridge collapse can be observed. The 30 year term is related to the engineering generation related to some particular concepts and tools.

The extension of this theory by Petroski (2006) in the 1990s assuming a catastrophic failure of cable stayed bridges has not yet been confirmed, however the vibration problems of the Millennium bridge in London have been associated with this theory. These considerations and the decades and centuries of discussion about the safety of bridges (Pugsley 1968; Jaeger 1970; Kurrer 2012; Duntemann & Subrizi 2000) and the evaluation of collapses (Tweed 1969) suggest that bridges are in a late stage of the technological development.

6.10 Conclusion

In this chapter various studies concerned with bridge failure and bridge failure statistics have been combined. The studies were combined regarding the collapse frequency and the collapse causes. In some studies, the collapse causes have been related to the age of the bridges, to the construction material, to the static systems and to other factors. Finally, the collapse frequencies have been compared to target values of the probability of failure and to the computed probabilities of failure.

The investigation has shown that the observed collapse frequencies have significantly decreased over the last two centuries. Current data indicates a cluster of bridge collapse frequencies in the range between 10^{-6} and 10^{-4} per year.

Furthermore it can be concluded that the major cause of bridge collapses are accidental events, mainly floods and impacts. Indirectly these causes may be related to insufficient knowledge or errors during the design process or due to changing service conditions of the bridges.

The estimation of the correct accidental loads remains a difficult issue. Proske (2015) has shown that many natural accidental loads were underestimated in the last century (floods, winds, earthquakes). Recent seismic hazard studies have confirmed that. The same trend can be seen for technical accidental loads such as impacts.

However, if only dead and live loads are considered, then bridges show an excellent behaviour yielding to bridge collapse frequencies in the range of 10^{-6} per year.

Additionally, it can be concluded that the collapse frequencies and the probabilities of failure fit very well. If we further consider correction factors for human failures, correlations and the time dependency of the target values of the probability of failure, the collapse frequencies comply even better and are covered by the target function.

Based on the last section we can conclude that

- observed collapse frequencies and target probabilities are in compliance,
- the safety target for bridges is fulfilled,
- the probabilities of failure and experience can both be equivalently used for the determination of design values and partial safety factors,
- the changes of the collapse frequencies due to limited maintenance is still to low to be observed in the collapse data,
- bridges are an extremely safe technical product and
- bridges are in a very late stage of the development cycle of technical processes. This late stage is usually characterised by high safety standards and sophisticated quality procedures.

News and media is regularly reporting about the decreasing standard of the bridge infrastructure and about collapsed bridges (Diehm and Hall 2013; Clifford Law Offices PC 2013; Nagel et al. 2016; Nachrichten 2017). Lawyers in the U.S. provide assistance for victims of bridge collapse accidents (Abels and Annes 2017). News reports about collapsed bridges have even been used for research regarding the social construction of risks (Stallings 1990).

However, all investigations carried out and conclusions drawn indicate that the decreasing standard is not (yet) visible in collapse and casualty numbers.

References

Abels & Annes PC (2017) Bridge collapse accidents

Akesson B (2008) Understanding bridge collapses. CRC Press, Taylor and Francis, London

Akiyama M, Frangopol DM, Arai M, Koshimura S (2012) Probabilistic assessment of structural performance of bridges under Tsunami hazard. In: 43rd Structures Congress, Chicago. 29–31 March 2012

Ang AH-S (2013) Minimizing the effects of uncertainty in life-cycle engineering. In: Strauss A, Frangopol DM, Bergmeister K (eds) Life-cycle and sustainability of civil infrastructure systems. Taylor & Francis Group, London, pp 36–41

Anonymus (2004) Nachrichten, der Prüfingenieur. April 2004, Seite 7

Anumba C, Egbu C, Kashyap M (2006) Avoiding structural collapses in refurbishment—a decision support system. Loughborough University, Prepared for the Health and Safety Executive, Research Report 463, Loughborough

ASCE (2013) 2013 Report card for America's infrastructure: bridges: overview, American society of civil engineers. http://www.infrastructurereportcard.org

BAST (2016) Zustandsnoten der Brücken. http://www.bast.de/DE/Statistik/Bruecken/Zustandsnoten.pdf?__blob=publicationFile&v=2

BAST (2017) Brücken an Bundesfernstrassen. http://www.bast.de/DE/Statistik/Bruecken/Brueckenstatistik.pdf?__blob=publicationFile&v=7

Biezma MV, Schanack F (2007) Collapse of steel bridges. J Perform Con-structed Facil (ASCE) 21(5):398–405

Binder F, Strauss A (2017) Life cycle costs for maintenance strategies using Monte Carlo simuation methods. In: Voigt M, Proske D, Graf W, Beer M, Häussler-Combe U, Voigt P (eds) Proceedings of the 15th international probabilistic workshop and the 10th Dresdener Probabilistik workshop, 27–29. Sept 2017. TUD press, Dresden, pp 97–108

Bolukbasi M, Mohammadi J, Arditi D (2004) Estimating the future condition of highway bridge components using national bridge inventory data, vol 9, no 1. Practical Periodical on Structural Design and Construction, ASCE, pp 16–24

Boysen-Hogrefe J, Jannsen N, Gern K-J, Groll D, Kappler M, Kooths S, Sachs A, Scheide J (2013) Mittelfristprojektion für Deutschland im Frühjahr 2013, Institut für Weltwirtschaft an der Universität Kiel, Prognose-Zentrum, 20 Mar 2013

Bozzolo D (1987) Ein mathematisches Modell zur Beschreibung der Dynamik von Steinschlag. ETH Zürich, Promotionsarbeit

Brady S (2013) The 30 year failure cycle. The Structural Engineer, May 2013, pp 14–15

Breysse D, Ndiaye A (2014) Failure case databases related to risk in civil engineering. In: Proceedings of the Institution of Civil Engineers—Forensic Engineering, Feb 2014, vol 167, issue 1, pp 27–37

Briaud JL, Gardoni P, Yao C (2012) Bridge Scour risk. In: ICSE6 conference proceedings, Paris, 27–31 Aug 2012, pp 1193–1210

Bridge Forum (2017) Bridge failure database. University of Cambridge. Available at: http://www.bridgeforum.org/dir/collapse/year/0000-3000.html

Brown CB (1979) A fuzzy safety measure. ASCE J Eng Mech Div 105:855–872

Brückenweb (2017) Brückenkatastrophen - Einstürze - Unfälle (Datenbank), F. Selke (Impressum). http://www.brueckenweb.de/2content/datenbank/katastrophen/3katastrophen.php

Budelmann H, Holst A, Wachsmann A (2013) Durability related life-cycle assessment of concrete structures: mechanisms, models, implementation. In: Strauss A, Frangopol DM, Bergmeister K (eds) Life-cycle and sustainability of civil infrastruture sytems. CRC Press, Taylor and Francis, London, pp 47–54

BMVBS (2013) Bericht Strategie zur Ertüchtigung der Straßenbrücken im Bestand der Bundesfernstrassen, Bundesministerium für Verkehr, Bau und Stadtentwicklung, 22 Mai 2013, Berlin

Cabinet Office (2015) National risk register of civil emergencies, London

Casas JR (2015) The bridges of the future or the future of bridges? Front Built Environ 1–3. https://doi.org/10.3389/fbuil.2015.00003
Christian GA (2010) Bridge Failures—Lessons learned. Bridge Engineering Course, University at Buffalo, 29 March 2010. http://mceer.buffalo.edu/education/bridge_speaker_series/2009-2010/presentations/P1%20Lessons%20learned%20from%20Bridge%20Failures_FINAL.pdf
Clifford Law Offices PC (2013) The shocking statistics on U.S. Bridge Collapses, 3 June 2013
Cluff LS (2007) Effects of the 2004 Sumatra-Andaman Earthquake and Indian Ocean Tsunami in Aceh Province. The Bridge, National Academy of Engineering, Washington, pp 12–16
Consolazio GR, Davidson MT, Getter DJ (2010) Vessel Crushing and Structural Collapse Relationships for Bridge Design, Aug 2010. University of Florida, Gainesville
Cook W (2014) Bridge Failure Rates, Consequences, and Predictive Trends. Dissertation, Utah State University Logan
Das PC (1997) Safety of Bridges. Thomas Telford, London
Davidson MT (2010) Probability Assessment of Bridge Collapse under Barge Collision Loads, Dissertation, University of Florida: Gainesville
Davis-Mcdaniel C (2011) Fault-tree model for bridge collapse risk analysis. M.Sc. Thesis, Clemson University
Deng L, Wang W, Yu Y (2016) State-of-the-art review on the cause and mechanisms of bridge collapse. J Perform Constructed Facil 30(2)
Diaz EEM, Moreno FN, Mohammadi J (2009) Investigation of common cause of bridge collapse in Colombia. Practice Periodical on Structural Design and Construction, vol 14, issue 4
Diehm J, Hall K (2013) Bridge collapses and structurally deficient bridges across the country. Huffington Post 24(5):2013
Dubbudu R (2016) An average of 7 structures collapsed per day in the last 5 years. Factly: Making Public Data Meaningful, 1 April 2016. https://factly.in/more-than-13000-lost-lives-in-structure-collapses-in-the-last-5-years/
Dunker KF (1993) Why America's Bridges are crumbling, Scientific American, vol 266, no 3, Mar 1993, pp 18–25
Duntemann JF, Subrizi CD (2000) Lessons learned from bridge construction failures. In: 2nd Forensic Engineering Congress, ASCE, Reston, VA, pp 374–385
Eberlin S, Hock B (2014) Zuverlässigkeit und Verfügbarkeit technischer Systeme – Eine Einführung in die Praxis. Springer Vieweg, Wiesbaden
Ellingwood B, Galambos TV, MacGregor JC, Cornell CA (1980) Development of a probabil-ity based load criteria for American National Standard A58, NBS Special Publication 577, U.S. Department of Commerce, National Bureau of Standard, Washington DC
Ellingwood B, Galambos TV, MacGregor JC, Cornell, CA (1982) Probability based load criteria: Load factors and load combinations. J Struct Div, ASCE 108(5): 978–997
Ellingwood BR (2001) Acceptable risk bases for design of structures. In: Progress in structural engineering and materials, vol 3, issue 2, Apr/June 2001, pp 170–179
Elm DG (1998) System health approach for risk management and design. In: Shiraishi, Shinozuka, Wen (eds) Structural safety and reliability. Balkema, Rotterdam, pp 271–277
Enright B, Caprani CC, O'Brien EJ (2011) Modelling of highway bridge traffic loading: some recent advances. In: 11th international conference on applications of statistics and probability in civil engineering (ICASP11), Zurich, 8 pages, on CD
Eurocode 0 (2017) EN 1990 basis of structural design, 2nd edn, Draft 30 April 2017
Fard B (2012) A comprehensive study on 100 bridge failures and their reduction strategies. Civil Engineering, University of Buffalo
FHWA (2017) National Bridge Inventory (NBI). https://www.fhwa.dot.gov/bridge/nbi.cfm
Fischer M (2010) Bestimmung modifizierter Teilsicherheitsbeiwerte zur semi-probabilistischen Bemessung von Stahlbetonkonstruktionen im Bestand, Dissertation, Kaiserslautern
Frandsen AG (1983) Accidents involving bridges. IABSE-reports 41:11–25

Frangopol DM, Okasha NM (2008) Life-cycle performance and redundancy of structures In: Graubner CA, Schmidt D, Proske D (eds) Proceedings of the 6th international probabilistic workshop, Darmstadt, pp 1–14

FSV (2017) Lebenszykluskostenermittlung für Brücken, RVS 13.05.11, Österreichische Forschungsgesellschaft Strasse – Schiene –Verkehr (FSV), Bundesministerium für Verkehr, Innovation und Technologie (bmvit), Vienna, 15 March 2017

Fu Z, Ji B, Cheng M, Maeno H (2012) Statistical analysis of the causes of bridge collapse in China. In: Sixth congress on forensic engineering

Gradner (2016) http://www.gartner.com/technology/research/methodologies/hype-cycle.jsp

Gucma L (2015) Risk management in the area of bridge situated on waterways in the aspect of ships collisions. Marine Traffic Engineering—MTE, Szczecin

Hamill L (1999) Bridge hydraulics. E & FN Spon, London

Hannawald F, Reintjes KH, Graße W (2003) Messwertgestützte Beurteilung des Gebrauchsverhaltens einer Stahlverbund-Autobahnbrücke. Stahlbau 72(7):507–516

Harik IE, Shaaban AM, Gesund H, Valli YS, Wang ST (1990) United States bridge failures, 1951–1988. J Perform Constructed Facil (ASCE) 4(4):272–277

Hersi M (2009) Analysis of bridge failure in United States (2000–2008). Ohio State University. Master of Science Thesis

Hong JH, Chiew YM, Lu JY, Lai JS, Lin YB (2012) Houfeng bridge failure in Taiwan. J Hydraulic Eng pp 186–198

Imam BM, Chryssanthopoulos MK (2012) Causes and consequences of metallic bridge failures. Struct Eng Int 22(1):93–98

Imhof D (2004) Risk assessment of existing bridge structures. University of Cam-bridge. Dissertation, Kings College

Jaeger TA (1970) Das Risikoproblem in der Technik. Schweizer Archiv 36:201–207

Jan C-D, Chen C-L (2005) Debris flows caused by Typhoon Herb in Taiwan. In: Jakob M, Hungr O (eds) Debris flow hazards and related phenomena. Springer, Berlin-Heidelberg, pp 539–563

Kosa J (2014) Damage analysis of bridges affected by the Tsunami in the Great East Japan earthquake. J JSCE 2:77–93

Kotes P, Vican J (2012) Reliability levels for existing bridges evaluation according to Eurocode. Procedia Eng 40:211–2016

Kraus U (2012) Lehren aus 2002 - Erfahrungen und Maßnahmen, Konferenz "Alle in einem Boot" - 10 Jahre transnationale Zusammenarbeit beim Hochwasserrisikomanagement an der Labe-Elbe, 21–22 June 2012, Präsentation

Kurrer KE (2012) The History of the Theory of Structures: From Arch Analysis to Computational Mechanics. Ernst & Sohn, Auflage 1

Lee GC, Mohan SB, Huang C, Fard BN (2013a) A study of U.S. bridge failures (1980–2012). Technical report MCEER-13-0008, 15 June 2013, University at Buffalo, State University of New York

Lee GC, Qi J, Huang C (2013b) Development of a database framework for modelling damaged bridges. Technical report MCEER-13-0009, 16 Jun 2013, University at Buffalo, State University of New York

Lehmann U (2003) Die Flutkatastrophe in Sachsen-hohe Schäden im ÖPNV. In: VDV Jahresbericht 2002–2003, Köln, pp 108–109

Main-Netz (2009) Bei Schiffskollision nicht alle standfest. www.main-netz.de/nachrichten/region/aschaffenburg/aschaffenburg-land/land/art3986.678394

Maruyama K, Tanaka Y, Hosoda A (2012) Damage of bridges structures by huge Tsunami and evaluation of Tsunami force on bridges. In: The 8th international symposium on social management systems, SSMS2012—Disaster prevention and reconstruction management. 2–4 May 2012, Kaohsiung, Taiwan

Mastaglio L (1997) Bridge bashing. Civil Eng pp 38–40

McLinn J (2010) Major bridge collapses in the US and around the world. IEEE Reliability society 2009 annual technology report. IEEE transactions on reliability, vol 59, heft 3, September 2010, 5 pages

McSaveny MJ, Davies TRH (2005) Engineering for debris flows in New Zealand. In: Jakob M, Hungr O (eds) Debris flow hazards and related phenomena. Springer, Berlin-Heidelberg, pp 635–658

Mende V (2016) Militärischer Einfluss auf Konstruktion und Architektur von Eisenbahnbrücken im Deutschen Reich, Tagungsband des 26. In: Dresdner Brückenbausymposiums, Technische Universität Dresden, Dresden, 14–15 Mar 2016, pp 221–233

Menzies JB (1996) Bridge failures, hazards and societal risk. In: International Symposium on the Safety of Bridges, July 1996, London

Metzner A (2002) Die Tücken der Objekte – Über die Risiken der Gesellschaft und ihre Wirklichkeit. Campus Verlag GmbH, Frankfurt Main

Michler H (2013) Innovativ! Leicht! Formbar! Bewährt! - Textilbetonbrücke Rottachsteg Kempten im Allgäu. In: Proceedings of the 23. Dresdner Brückenbausymposium, Technische Universität Dresden, Dresden, pp 239–252

Michler H (2016) Verstärken mit Carbonbeton im Brückenbau, 26. In: Dresdner Brückenbausymposium, Technische Universität Dresden, 14–15 Mar 2016, pp 235–248

MLIT (2005) The maintenance of national road network in Japan. Ministry of land, infrastructure, transport and tourism. http://www.mlit.go.jp/road/road_e/03key_challenges/1-2-2.pdf

Montalvo C, Cook W (2017) Validating common collapse conjectures in the U.S. bridges. In: 11th international bridge and structure management conference, 25–27 April 2017, Mesa, Arizona, 13 pages

Morrissey MM, Savage WZ, Wieczorek GF (1999) Air blasts generated by rockfall impacts: analysis of the 1996 Happy Isles event in Yosemite National Park. J Geophys Res 104(B10):23189–23198

Nachrichten S (2017) Bröckelnde Bahnbrücken – Mehr als 100 Bauwerke im Land marode, 27 May 2017

Nagel L-M, Pauly M, Mucha V, Setzer J, Wilhelm F (2016) Wettlauf gegen den Verfall, Die Welt. http://www.welt.de/politik/interaktiv/bruecken/deutschlands-bruecken-wettlauf-gegen-den-verfall.html, https://docs.google.com/spreadsheets/d/1h_NKP3lvTnoQBFjCRa9gJLLOi2rxtIMeNucvAUsPwtI/edit?pref=2&pli=1#gid=0

Naumann J (2002) Aktuelle Schwerpunkte im Brückenbau, 12. Dresdner Brückenbausymposium, Lehrstuhl für Massivbau und Freunde des Bauingenieurwesens e.V., 2002, pp 43–5

Naumann J (2004) Bauwerksprüfung nach DIN 1076 - Bedeutung, Verantwortung, Durchführung, 14 Dresdner Brückenbausymposium. TU Dresden und Verein der Freunde des Bauingenieurwesens, Dresden, pp 53–64

Nendo S, Niczyj J (1998) Reliability estimation of trusses by using fuzzy set. In: Shiraishi, Shinozuka & Wen (eds) Structural safety and reliability. Balkema, Rotterdam, pp 317–324

NIST (2006) Performance of physical structures in Hurricane Katrina and Hurricane Rita: a reconnaissance report, NIST technical note 1476. National Institute of Standards and Technology, Gaithersburg

Nowak AS, Collins KR (2012) Reliability of structures, 2nd edn. CRC Press, Boca Raton, FL

Oberländer B (2012) Brücke über den Sundgraben hängt in den Seilen, BZ Berner Zeitung. http://www.bernerzeitung.ch/region/thun/Bruecke-ueber-den-Sundgraben–haengt-in-den-Seilen-/story/27896453. Accessed 24 May 2012

Petroski H (2006) Success through Failure: The Paradox of Design. Princeton University Press, Princeton

Petschacher M (2007) Stochastisches Alterungsmodell für Infrastrukturbauten - Untersuchung des Brückenbestandes im hochrangigen Strassennetz. PEC - Petschacher Consulting, ZT GmbH, Feldkirch

Pircher M, Kammersberger A, Lechner B, Mariani O (2009) Schädigung einer schlaff bewehrten Betonbrücke durch Verkehrsbelastung. Beton- und Stahlbetonbau 104(3):154–163

Proske D (2004) Beitrag zur Risikobewertung von alten Brücken bei Schiffsanprall. Dissertation, Technisch Universität Dresden, Dresden

Proske D (2015) The uncertainty of the estimation of natural life loads. Bautechnik 92(12):854–859

Proske D, van Gelder P (2009) Safety of historical stone arch bridges. Springer, Berlin—Heidelberg

Pugsley A (1968) The safety of bridges. The Struct Eng 46:197

Raithel, M, Leusink E, Franke H-J, Haupt R (2011) Geotechnische Problemstellungen bei der Sicherung bestehender Mainbrücken gegen Schiffstoß. Geotechnik-Tag in München 2011, 4 März 2011

Reichelt A, Richter S (2003) Hochwasserschäden im Vorlandbereich der Elbebrücke Riesa - Maßnahmen zu deren Beseitigung, 13. Dresdner Brückenbausymposium, TU Dresden und der Verein der Freunde des Bauingenieurwesens, pp 157–187

Scheer J (2010) Failed bridges–Case studies, Causes and Consequences. Wilhelm Ernst & Sohn, Berlin

Schneider R, Fischer J, Bügler M, Nowak M, Thöns S, Borrmann A, Straub D (2015) Assessing and updating the reliability of concrete bridges subjected to spatial deterioration—principles and software implementation. Struct Concrete 16(3):356–365

Seipelt D, Eschweiler S, Neysters T, Coors B, Grassl M (2016) Reparatur der Autobahnbrücke über die Süderelbbrücke nach schwerem Schiffsanprall - Nachrechnung, Planung, Ausführung, Analyse, 26. Dresdner Brückenbausymposum, Technische Universität Dresden & Verein der Freunde des Bauingenieurwesens, pp 165–176

Sharma S (2010) A comprehensive study on bridge failures in the United States and their comparison with bridge failure in other countries. M.Sc. Project, Department of Civil, Structural and Environmental Engineering, University of Buffalo, Buffalo

Sharma S, Mohan S (2011) Status of bridge failures in the United States (1800–2009). TRB 90th annual meeting: transportation, liveability, and economic development in a changing World, Washington D.C.

Shoji G, Moriyama T (2007) Evaluation of the structural fragility of a bridge structure subjected to a Tsunami wave load. J Nat Disaster Sci 29:73–81

Shroder JF (2014) Natural resources in Afghanistan: geographic and geologic perspectives on centuries of conflict. Elsevier, San Diego

SIA 269 (2007) Grundlagen der Erhaltung von Tragwerken; Schweizerischer Ingenieur-und Architektenverein, Zürich, Draft 03/2007

Sibly PG, Walker AC (1977) Structural accidents and their causes. Proceedings of the Institution of Civil Engineers, 62(2):191–208

Simandl T, Glatzl J, Schweighofer B, Blovsky S (2006) Schiffsanprall an die Eisenbahnbrücke in Krems. Erstmalige Anwendung des neuen Eurocode EN 1991-1-7 für Außergewöhnliche Einwirkungen. Beton- und Stahlbetonbau 101(9):722–728

Smith DW (1976) Bridge failures. In: Proceedings of the institution of civil engineers, vol 60, issue 3, August 1976, pp 367–382

Sornette D (2000) Critical phenomena in natural sciences, chaoss, fractals, selforganization and disorder: concepts and tools. Springer-Verlag, Berlin—Heidelberg

Spaethe G (1992) Die Sicherheit tragender Baukonstruktionen, zweite, neubearbeite Auflage. Springer, Wien

Stallings RA (1990) Media discourse and the social construction of risk. Soc Probl 37(1):80–95

Starr C (1969) Social benefit versus technological risk. Science 165:1232–1238

Steedman S (2010) The Long Learningcurve, Ingenia, Issue 44, September 2010, pp 3

Stulc J (2015) The 2002 floods in the Czech Republic and their impact on built heritage, pp 133–137

Taricska MR (2014) An analysis of recent bridge failures in the United States (2000–2012). M.Sc. Thesis, The Ohio State University

Tweed MH (1969) A summary and analysis of bridge failures. Master Thesis, Iowa State Uni-versity, Ames

Unjoh S (2005) Damage to transportation facilities, the damage induced by Sumatra earthquake and associated Tsunami of 26 Dec 2004. A report of the reconnaissance team of Japan society of civil engineers, pp 66–76

Van Lente H, Spitters Ch, Peine A (2011) Comparing technological hype cycles: towards a theory. ISU working paper 11.03, Innovation Studies Utrecht (ISU), Utrecht University, Copernicus Institute of Sustainable Development

Vogel T, Zwicky D, Joray D, Diggelmann M, Hoj NP (2009) Tragsicherheit der bestehenden Kunstbauten, Sicherheit des Verkehrssystems Strasse und dessen Kunstbauten, Bundesamt für Strassen, Dezember 2009, Bern

Vollrath F, Tathoff H (2002) Handbuch der Brückeninstandhaltung. VBT Verlag Bau + Technik, Düsseldorf

von Kirchbach H-P, Franke S, Biele H (2002) Bericht der Unabhängigen Kommission der Sächsischen Staatsregierung Flutkatastrophe 2002, Dresden

Wang N (2010) Reliability-Based Condition Assessment of Existing Highway Bridges. Dissertation, Georgia Institute of Technology, August 2010

Wardhana K, Hadipriono FC (2003) Analysis of recent bridge failures in the United States. J Perform Constructed Facil ASCE 144–150

Weber WL (1999) Die gewölbte Eisenbahnbrücke mit einer Öffnung. Dissertation, Techni-sche Universität München, München

Wei X, Wang Q, Wang J (2008) Damage pattern and failure mechanisms of bridge pile foundation under earthquake. In: Proceedings of the 14th world conference on earthquake engineering, 12–17 October 2008, on CD, Beijing, China, 7p

Wenk T (2005) Beurteilung der Erdbebensicherheit bestehender Straßenbrücken. Bundesamt für Strassen ASTRA, Bern

Wikipedia (2017) List of bridge failures. https://en.wikipedia.org/wiki/List_of_bridge_failures

Woodward RJ, Kaschner R, Cremona C, Cullington D (1999) Review of current procedures for assessing load carrying capacity—status C, BRIME PL97-2220, Jan 1999

Xu FY, Zhang MJ, Wang L, Zhang JR (2016) Recent highway bridge collapses in China: review and discussion. J Perform Constructed Facil 30(5) (Oct 2016), 8p

Yan BF, Shao XD (2008) Application of China bridge management system in Qinyuan city. In: Koh HM, Frangopol DM (eds) Bridge maintenance, safety, management, health monitoring and informatics. Taylor & Francis Group, London, pp 2675–2682

Yim SC, Azadbakht M (2013) Tsunami forces on selected California coastal bridges. School of Civil & Construction Engineering, Oregon State University. Final report No. CA13-1983

Zerna W (1983) Grundlage der gegenwärtigen Sicherheitspraxis in der Bautechnik. In: Hartwig S (ed) Große technische Gefahrenpotentiale: Risikoanalysen und Sicherheitsfragen. Springer Verlag, Berlin—Heidelberg, pp 99–109

Zhang S (1993) A comprehensive approach to the observation and prevention of debris flow in China. Nat Hazards 7:1–23

Zhang XG, Liu G, Ma JH, Wu HB, Wu WM (2014) Design concept and approach on sustainable development of bridge engineering. In: Chen A, Frangopol DM, Ruan X (eds) Bridge maintenance, safety, management and life extension. CRC Press, London, pp 1831–1838

Zhang J, Chen X, Liu D, Li X (2016) Analysis of bridge response to barge collision: Refined impact force models and some new insights. Adv Struct Eng 19(8):1–21

Zorn M, Komac B (2008) The debris flow in Log pod Mangartom, NW Slovenia, WIT transactions on engineering sciences, vol 60. In: Wrachien, Aristide, Brebbia (eds) Monitoring, simulation, prevention and remediation of dense debris flows II. WIT Press, pp 125–133

Chapter 7
Conclusion

In this book it was shown that the values of computed probabilities of failure of bridges and the observed frequencies of collapse of bridges are comparable by using correction factors for example for human failures and correlation values. Although two factors can easily be applied to fit one target parameter to another, major efforts have been undertaken to justify all parameters. Information is given for almost all of the parameters.

Additionally the book has clearly identified the major causes of the collapse of bridges: accidental loads. These loads could be managed adequately in the codes of practice as we have seen in the last decades for seismic loading. As an example Fig. 7.1 shows the distribution of damages on bridges caused by earthquakes related to the generations of the design codes used for the bridge. It can be clearly seen that the damages are significantly lower for bridges based on new codes with stricter requirements. For this conclusion it was also shown that the age of the bridges has only a minor effect.

On the other hand these loads do not also affect the stability of the bridges itself, they also affect the conclusion in this book since their large return periods limit the conclusions drawn from collapse data with short time periods.

In several cases the responsible authority for the selection of the design loads of such accidental impact loads is the same authority which is financially responsible for the strengthening of existing structures for new and updated loads. For example if train impact loads are suggested by the railway companies, there is a semblance of prejudice. We do not find such a semblance of prejudice for example for seismic loading since the government requires simply stronger rules which have to be implemented by the individual investors as well as organisations.

Another issue is the limited exposure of the public to such causes of collapse. We are usually exposed to the risk of bridge failure not more than a few minutes per day leaving aside possible traffic jams on bridges. Therefore the risk for the public is limited. During some accidental loads further mitigation actions are taken, for example the safety of the public during floods is often provided by limited access or closure of bridges. Figure 7.2 shows the Blue Wonder Bridge in Dresden. This bridge

D. Proske, *Bridge Collapse Frequencies versus Failure Probabilities*, Risk Engineering,
https://doi.org/10.1007/978-3-319-73833-8_7

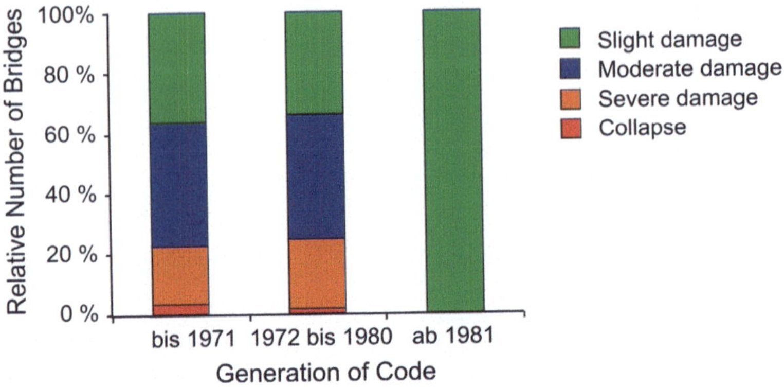

Fig. 7.1 Seismic-induced damages on bridges during the Northridge earthquake in the area of Los Angeles with regard on the code generation for the bridge design (Wenk 2005)

Fig. 7.2 Blue wonder bridge in Dresden (*Picture* D. Proske)

is closed for heavy traffic during large floods since the embankment can be flooded too which affects the load bearing capacity of the bridge. Therefore bridges may collapse, however humans are not directly affected by the collapse of these bridges. This conclusion is in contrast to the collapses related to impacts such as the collapses of bridges due to ship impact and the collapses of bridges due to train impact.

The conclusion of the book is that major causes of the collapse of bridges are manageable from a technical point of view, there is no sudden surprise in the data. The probabilistic safety concept is confirmed by the observed collapse frequencies of bridges. The risk of human loss related to loads and causes is mostly limited except for the impacts. Here we have seen disasters over the last decades with more than 200 fatalities. Therefore further improvements are recommended.

Changes in the loads related to climate change or changed maintenance strategies are not yet visible in the collapse frequency data.

Reference

Wenk T (2005) Beurteilung der Erdbebensicherheit bestehender Straßenbrücken. Bundesamt für Strassen ASTRA, Bern

Index

D. Proske, *Bridge Collapse Frequencies versus Failure Probabilities*, Risk Engineering,
https://doi.org/10.1007/978-3-319-73833-8

The manufacturer's authorised representative in the EU is Springer Nature Customer Service Centre GmbH, Europaplatz 3, 69115 Heidelberg, Germany. If you have any concerns regarding our products, please contact ProductSafety@springernature.com

Printed and bound by CPI Group (UK) Ltd, Croydon, CR0 4YY
22/09/2026
02220990-0008